Praise for *The Wisdom Paradox*

"This is an optimistic book. . . . There is even some evidence that a positive mental attitude can help ward off cognitive decline, in which case it might be worth reading the book for its cognitive-enhancement properties alone." —Laura Spinney, *New Scientist* (UK)

"Goldberg observes that exposure to similar, new things creates neural networks in the brain that attract each other and accumulate, networks that in some circumstances are expressed as expertise and in others as intuition (or both). The networks accrue with age—Goldberg ventures to call the result of this accumulation wisdom." —Sue M. Halpern, *The New York Review of Books*

"[Goldberg] has found that our intuitive powers grow stronger with every passing year. . . . There is now hope that we can be as mentally alert—if not more so—in our twilight years than we have ever been." —Jessica Kiddle, *The Scotsman* (UK)

"I am now spoiled; I need more essays by opinionated, original, and intellectual contemporary scientists."
—Nassim Nicholas Taleb, author of *Fooled by Randomness*

"A book of wise reflections on the gains, not the losses, that come to the older human mind. Here is a valuable addition to the literature on aging."
—Antonio Damasio, author of *Descartes' Error, The Feeling of What Happens*, and *Looking for Spinoza*

"A refreshing reprieve from the onslaught of negatives about the aging process. It is both scholarly and a pleasure to read, filled with a multitude of cogent observations on how a host of human endeavors benefit from aging, often in the face of degenerative disease. It is destined to inspire more of our aging population to strive for greatness."
—Georg Deutsch, coauthor of *Left Brain, Right Brain*

"Goldberg has done a marvelous job in distinguishing between disease and cognitive decline that occurs in normal aging. It is a must-read for people who are interested in preserving and improving their brains." —Horace Deets, former executive director of AARP

"*The Wisdom Paradox* is a delight to read, a splendid review of modern neuroscience. The book is erudite, challenging, and above all, hopeful. I recommend it to all who are interested in keeping their wits about them."

—Allan Mirsky, chief, section on clinical and experimental neuropsychology, National Institute of Mental Health

SERGEY KNAZEV AND PETER V. LANG

Elkhonon Goldberg is the author of *The Executive Brain* and a clinical professor of neurology at New York University School of Medicine. He divides his time between private practice in neuropsychology, research in cognitive neuroscience, and teaching. He lives in New York City.

THE WISDOM PARADOX

ALSO BY ELKHONON GOLDBERG

The Executive Brain: Frontal Lobes and the Civilized Mind

The WISDOM PARADOX

How Your Mind Can Grow Stronger As Your Brain Grows Older

ELKHONON GOLDBERG

GOTHAM BOOKS

GOTHAM BOOKS
Published by Penguin Group (USA) Inc.
375 Hudson Street, New York, New York 10014, U.S.A.

Penguin Group (Canada), 90 Eglinton Avenue East, Suite 700, Toronto, Ontario, Canada M4P 2Y3 (a division of Pearson Penguin Canada Inc.); Penguin Books Ltd, 80 Strand, London WC2R 0RL, England; Penguin Ireland, 25 St Stephen's Green, Dublin 2, Ireland (a division of Penguin Books Ltd); Penguin Group (Australia), 250 Camberwell Road, Camberwell, Victoria 3124, Australia (a division of Pearson Australia Group Pty Ltd); Penguin Books India Pvt Ltd, 11 Community Centre, Panchsheel Park, New Delhi - 110 017, India; Penguin Group (NZ), Cnr Airborne and Rosedale Roads, Albany, Auckland, New Zealand (a division of Pearson New Zealand Ltd); Penguin Books (South Africa) (Pty) Ltd, 24 Sturdee Avenue, Rosebank, Johannesburg 2196, South Africa

Penguin Books Ltd, Registered Offices: 80 Strand, London WC2R 0RL, England

Published by Gotham Books, a division of Penguin Group (USA) Inc.
Previously published as a Gotham Books hardcover edition.

First trade paperback printing, February 2006

10 9 8 7 6 5 4 3 2 1

The Library of Congress has cataloged the hardcover edition of this title as follows:

Goldberg, Elkhonon.
The widsom paradox : how your mind can grow stronger as your brain grows older / Elkhonon Goldberg.
p. cm.
ISBN 1-592-40110-4 (hardcover) ISBN 1-592-40187-2
1. Brain—Aging. 2. Neuropsychology. 3. Cognitive neuroscience. I. Title.
QP356.25.G64 2005
612.8'2—dc22 2004060695

Printed in the United States of America
Set in Bembo
Designed by Susan Hood

TO MY FELLOW BABY BOOMERS—
THE HEADSTRONG GENERATION

CONTENTS

"WISDOM BEGINS IN WONDER."

—SOCRATES

THE WISDOM PARADOX

INTRODUCTION

Ruminations of a Baby Boomer Neuroscientist

Like Tolstoy's unhappy families in *Anna Karenina*, the midlife crisis takes many forms. I first knew that mine was encroaching when halfway through my sixth decade, I began to search for a cathartic experience. An odd sense of temporal symmetry set in. For the first time in my life the past seemed as important as the future, and I felt an urge to examine this more deeply. I felt a sudden need to take stock of my life and to connect its pieces, disjointed by circumstance. For the first time in twenty-six years I visited the country of my birth to seek out old friends with whom I had had no contact in half a lifetime. And I wrote a book, an intellectual memoir of sorts, trying to place my past, my present, and my premonition of the future into a single coherent perspective.

For reasons existential more than immediate or practical, I also decided to take stock of the physical damage wrought by time. After many years of blatant self-neglect, I had a long overdue comprehensive physical exam. I was delighted to discover that by all objective medical criteria I was in good health, biologically younger than my chronological age. This pleased me but did not particularly surprise me, since I felt fine and my energy level had been undiminished with age.

With considerable trepidation, I also decided to have an MRI of my brain, a magnetic resonance imaging procedure to visualize the structures inside my aging skull. I had no indication that my mind was beginning to fail me. Quite the contrary, I had good reasons to believe that my cognition was fine: I had just published a reasonably successful book. I was lecturing worldwide and continued to get away with tackling arcane technical matters before demanding audiences without notes. At any given time, I was engaged in a number of parallel activities, usually without dropping the ball. My mental life was rich and full. My private practice in neuropsychology was booming and my career flourishing. And I took occasional mischievous pleasure in teasing my much younger assistants and graduate students that I had more physical stamina and mental focus than they did.

At the same time, I knew that I carried certain genetic baggage. There was no known history of dementia on either side of my family, but my mother had died of a stroke, albeit at the enviable age of ninety-five, and her younger brother, while basically of sound mind, had suffered from a relatively advanced brain vessel condition known as *multiinfarct disease*. I knew this because I had been the one who diagnosed his condition by reviewing the MRI of his brain.

More to the point, for many years my lifestyle had followed a rather unhealthy pattern. I grew up in Russia (the former Soviet Union, to be precise) and came to the United States at the age of twenty-seven. Having rejected my old country's political system, I continued to embrace many aspects of its self-destructive lifestyle. I chain-smoked from my teens until my early forties, when I kicked the habit finally and irreversibly, and for years I drank considerably more than is common among the middle-aged Jewish intellectuals on this side of the Atlantic. In short, I had plenty of neurotoxins in my background to answer for.

As a cognitive neuroscientist, I am used to regarding the brain dispassionately and abstractly in the laboratory. As a clinical neuropsychologist, I am trained to be exquisitely perceptive about the minute manifestations of brain dysfunction and brain damage—that is, other people's brain damage. The flip side of having the MRI was that I would be wrenchingly aware of any potential consequences of the condition of my own brain, and the prospect of gaining this knowledge scared me.

The paradox was not unique to me. In occasional conversations with more than a few friends—world-renowned neuroscientists, neurologists, and psychiatrists among them—they have all said that their curiosity about the condition of their own bodies stopped at the level of the neck. What was in their heads, they simply did not care to know. This agnostic disclaimer was invariably accompanied by a neurotic chuckle, and I could understand why.

But for me, uncertainty is usually a source of anxiety, while clarity, whatever its content, has always had a mobilizing effect. Among the assorted and often unflattering zoological appellations used by my friends and foes alike to capture my central personality traits, *ostrich* has never been invoked. I have always prided myself of being a reasonably courageous, head-on type, and now my head was about to be inserted into the brain scanner's magnetic coil. My neurosurgeon friend Jim Hughes, whom I asked for an MRI referral, first ridiculed the idea and tried to talk me out of it.

"What if we find a benign tumor?" Jim kept saying. "Your life will be ruined by torment!" He brought up the case of Harvey Cushing, arguably the father of American neurosurgery, who himself had a benign brain tumor.

To that I fatuously replied that surely I had enough character and inner strength to deal with any such findings rationally, and that, anyway, knowledge was better than ignorance.

"In that case, *my* life will be ruined by torment if we find something bad in *your* brain," Jim said in exasperation.

After some argument, we resolved that having Jim's life ruined by torment was an acceptable price to pay for satisfying my morbid curiosity, and Jim acquiesced.

As a clinical neuropsychologist and a cognitive neuroscientist, I have been studying the effects of various forms of brain damage on the human mind for thirty-five years, and I have seen and analyzed hundreds of CT and MRI brain scans. For the first time, however, I was about to see the images of my own brain. I knew better than most people how devastating even mild brain damage could be for the mind, and for the soul as well. But in the final analysis I meant every word that I had said to Jim. I believed that I could deal with any news, including bad news, and that knowledge was preferable to ignorance under any circumstances. So on a sunny April day, I walked into the offices of Columbus Circle MRI in midtown Manhattan.

The report and the films (not usually released to patients but released to me as a colleague) arrived a few days later. What I saw did not look terrible, but it did not particularly please me either. My cortical sulci (the walnut-shaped convolutions on the surface of the brain) and ventricles (spaces inside the brain containing the cerebrospinal fluid that bathes the brain) were declared by the radiologist to be "normal in size." By my own reckoning, the sulci unequivocally were, but the ventricles looked large to me even allowing for the expected, normal dilation (the technical term for enlargement) with age. This suggested some brain atrophy.

Furthermore, two tiny areas of increased signal intensity in the white matter (long nerve pathways connecting distant parts of the brain and encased in white fatty tissue called myelin) of the left hemisphere were noted in the report. I could see them also. The meaning of such findings is uncertain. In my case, they most likely reflected ischemic changes, regional death of brain tissue due to poor oxygen supply. They could also mean the loss of myelin in certain areas—probably a less likely explanation. By my own definition of the term, I had mild brain damage.

The news was not all bad. "Normal flow voids" were present in my internal carotid and basilar arteries, and diffusion images were unremarkable. This meant that my major arteries were clean as a whistle, not occluded, not cluttered with fatty debris, and that my blood vessels were strong. This was consistent with a normal ultrasound Doppler test of my carotid arteries, which I had had as part of my physical a few months earlier. Taken together with my somewhat high but generally normal blood pressure, these findings made the possibility of a sudden, major, catastrophic stroke or aneurysm rupture mercifully remote. The hippocampi (seahorse-shaped brain structures known to be important for memory) appeared normal in size—definitely a good thing, since hippocampal atrophy is a common harbinger of Alzheimer's disease.

To lay my apprehensions to rest, I paid a visit to one of New York's top neurologists, Dr. John Caronna at the famed New York Presbyterian Hospital (where many years ago, barely off the immigrants' boat, I held my first faculty position in the United States). Dr. Caronna, a genial and gregarious man, examined me carefully, looked at my scans, and showed them to a colleague, the head of neuroradiology at the Cornell University's Weil Medical School. They both concluded that everything was normal for my age, including the two "punctate" (a fancy way of saying "tiny") areas of ischemia.

"It's just a well-used brain, that's all," said Caronna with his characteristically endearing sense of humor.

Having seen hundreds of scans myself, however, I still felt that my ventricles were larger than those of many other people my age and that the tiny ischemic lesions apparent on my scan were not a sine qua non of aging. To resolve the issue, I showed the scans to an old friend, Dr. Sanford Antin. Sandy is among the most experienced neuroradiologists in New York, and I had collaborated with him in the past on some of the most formative projects of my scientific career.

Sandy looked at the MRI scans, immediately dismissing one of the two punctate lesions as an artifact, confidently and at length explaining to me how such artifacts happen. He then declared the other punctate lesion "insignificant," pronounced the sulci and gyri (tiny canyons between the sulci) "normal for any age," and complimented me on my "beautiful brain."

So, I was finally relieved of my personal apprehensions. In retrospect, I found my brain-scanning exercise interesting in two respects: both neurological and neurotic. Neurologically and neuropsychologically speaking, an argument can be made that what I did should become part of a routine physical checkup for people past a certain age, maybe not every year but perhaps every three to five years. We all recognize the utility of prophylactic tests, as well as the fact that our vulnerability to a whole range of afflictions increases with age. Hence the universal acceptance, in fact promotion by the medical establishment, of colonoscopy as a means of combating colon cancer, breast and prostate cancer tests, and so on. But the brain has been traditionally exempt from this prophylactic scrutiny, as if the brain was not of the body. This seems highly illogical, since the incidence of dementia in the aging population rivals, and often exceeds, the incidence of many other afflictions.

The Mind, the Brain, and the Body

Such an illogical and unfortunate state of affairs is probably predicated on two tacit assumptions, one coming from the general public and the other from health professionals. Until recently, the mind was not regarded by most people as part of one's biological being, subject to medical and quasimedical scrutiny. This, of course, is a misconception, an enduring legacy of Cartesian mind-body dualism. Today the general educated public is increasingly at home with the understanding that the mind is of the brain, and thus of the body. This will be one of the major themes of this book.

In the eyes of health professionals, the utility of an early diagnosis of the potentially dementing diseases of the brain is often doubted on the grounds that "nothing can be done about it anyway." To put it in military terms, this type of information is not deemed to be "actionable" and therefore is not useful, merely upsetting to the patient, and the diagnosis without treatment would merely place undue financial burden on society. This tacit and sometimes not-so-tacit assumption, so sadly accurate even a decade ago, is rapidly becoming obsolete, owing to the rapid advent of various pharmacological and nonpharmacological ways of protecting the brain against decay. In plain terms, the assumption that "nothing can be done" is no longer true.

But all the rational argument notwithstanding, I recognize that what I did was an exercise in neurotic behavior first and foremost. I am sure that a neurotic response to aging is common among millions of my contemporaries, no matter how enlightened (and perhaps the more enlightened the more so). It may take many forms. Being a neuroscientist, I immediately ordered a brain MRI. Others deal with their age neuroses in different ways. Often the neurosis takes the form of denial or, to put it more precisely, a refusal to know, which I have witnessed in several colleagues.

The whole experience provided a point of departure for some serious thinking about the fate of an aging mind in an aging brain in modern society. Like most things in life and in nature, brain health versus brain damage is not a simple binary distinction. There are shades of gray . . . even when it comes to the gray matter, so to speak.

The expression "baby boom" has a distinctly American ring to it, but the phenomenon itself is universal. During the decade following the end of World War II, birth rates exploded in Europe and in Russia, just as they did in North America. Today, in societies increasingly abuzz with concerns about "Alzheimer's epidemics," my anxieties are shared by millions of my enlightened contemporaries worldwide. Many of them, perhaps most, carry some baggage similar to mine in one form or another. What in their anxiety is neurotic, and what is justified? Part reality, part neurosis, a certain amount of anxiety about the state of one's mental faculties is common and expected in anyone approaching the "ripe middle age." In my case, this state of mind was colored, for better and for worse, by my professional knowledge of how the brain works, and of the many ways in which the brain may fail to work. I am different from most of my worried contemporaries in that I am a brain scientist and clinician diagnosing and treating various effects of brain damage for a living, dealing with aging minds and with dementia on a daily basis. This may make my insights into my own anxieties particularly useful to other people. And so I hope that the ruminations of an aging neuroscientist will be informative and useful to my aging contemporaries from every walk of life.

As young people, we are driven by the lust for the unknown, for forward motion. We *dare*. The folkloric cliché is that as we age, we yearn for stability. Does "stability" inevitably equal "stagnation"? Are age-associated mental changes all losses, or are there also some gains? As I am surveying introspectively my own mental landscape, I conclude that, despite my anxieties and increasingly precarious epidemiological odds, things are not all bad. I notice, with some satisfaction, that on balance I am no more stupid, in some intuitive sense, than I was thirty years ago. My mind is not dimmed; in some ways it may in fact be working better. And as psychological (and hopefully also real) protection against the effects of aging, I find myself constantly propelling myself into forward motion. I wage an unending internal war on stasis. A life too settled is no longer a life but an afterlife, and I want no part of it for myself.

What strikes me most in this introspective pursuit is that if there is a change, it cannot be captured in quantitative comparisons. On balance, my mind is neither weaker nor stronger than it was decades ago. It is different. What used to be the subject of involved problem-solving has become more akin to pattern recognition. I am not nearly as good at laborious, grinding, focused mental computations; but then again I do not experience the need to resort to them nearly as often. In my early twenties, I took pride (somewhat flippantly) in being able to follow a lecture on an arcane topic in advanced mathematics without taking notes, and to pass a test a few months later. I will not even attempt this feat at my ripe age of fifty-seven. It's simply too hard!

But other things have become easier. Something rather intriguing is happening in my mind that did not happen in the past. Frequently, when I am faced with what would appear from the outside to be a challenging problem, the grinding mental computation is somehow circumvented, rendered, as if by magic, unnecessary. The solution comes effortlessly, seamlessly, seemingly by itself. What I have lost with age in my capacity for hard mental work, I seem to have gained in my capacity for instantaneous, almost unfairly easy insight.

And another interesting bit of introspection: As I am trying to solve a thorny problem, a seemingly distant association often pops up like a deus ex machina, unrelated at first glance but in the end offering a marvelously effective solution to the problem at hand. Things that in the past were separate now reveal their connections. This, too, happens effortlessly, by itself, while I experience myself more as a passive recipient of a mental windfall than as an active, straining agent of my mental life. I have always strived to reach across the boundaries of professional and intellectual domains, but now, as this "pop-up" phenomenon is happening more often, I am finding this "mental magic" productive and incredibly satisfying—like a kid who finds a hidden cookie jar and helps himself with impunity and glee.

Then there is something else, even more profound, almost too good to admit: the feeling of being in control of my life, like I had never experienced before. At the risk of sounding hypomanic (I am not, which is why I feel free to say it), there is increasingly the feeling that life is a feast, when in the past the prevailing feeling had often been that life was a struggle. And despite the full awareness of the biological imperative that the feast will sometime come to an end, or maybe precisely because of this awareness, an urge swells, powerful like a force of nature and getting more powerful with age, to prolong the feast. The existential paradox of aging—to marvel about its effects and yet follow the drive for prolonging the feast. Because life is not a one-way street of decay. There are both currents and counter-currents to be lived, examined, understood, and enjoyed.

What are they, these strange phenomena of mental levitation, when solutions come instantly and without apparent effort? Is it, perchance, that coveted attribute of aging, that stuff of sages called *wisdom*? At first I feared getting carried away, lest my foray into the mysteries of wisdom prove to be an exercise in foolishness. I sought to stay away from such expansive poetic language and stick to the austere language of science, which has been my language most of my life, to speak not of "wisdom" but of "pattern recognition."

Yet as I caution myself against making extravagant claims, I find myself inexorably tempted by them, and the existential paradox, which intrigues me so, gradually takes on a new name: the *wisdom paradox*. Our minds are a function of the natural organism that is the brain. And though the brain may age and change, each phase of this progression presents new and different pleasures and advantages, as well as losses and trade-offs, in a natural progression, like the seasons. If our mental seeds are sown through curiosity and exploration early in youth, and experience in more mature life tends and nurtures the mental crop, then wisdom is the harvest of mental rewards that we can only truly enjoy in what Frank Sinatra famously called "the autumn of the years." And having taken a deep breath, I am plunging head-on into my new project, *this* project, a book about the seasons of a human mind as the passage from daring to wisdom. As I am embarking on my project, the thought cannot escape me that wisdom, with its cognitive, ethical, and existential dimensions, is far too rich a concept to be explored in its entirety in a single narrative, or by a single explorer such as myself. So I am deliberately limiting the scope of this book to the cognitive dimension of wisdom—a perspective that is admittedly narrow but eminently worthy of exploration nonetheless.

Book Overview

The multifaceted nature of the subject is reflected in the book's eclectic content and interweaving themes. In the narrative to follow, certain chapters focus on history and culture (chapters three, four, five, and twelve); others focus on psychology (chapters one, four, five, eight, nine, ten, eleven, and twelve); yet others focus on the somewhat more technical matters of how the brain is wired and how it functions and malfunctions (chapters two, six, seven, thirteen, and fourteen). Finally, I talk about what can be done to forestall the aging of the brain (chapters fourteen, fifteen, and the epilogue). These seemingly disparate subjects are united in a coherent logical thread driven by the central questions: What enables the aging brain to accomplish remarkable mental feats, and how can we enhance this ability? All the names of my patients are disguised to protect their privacy, but their stories are authentic and unembellished. I did my best to explain all the technical terms in the text where they first appear.

We will start with a casual walk through the not-so-casual brain machinery powering the seemingly mundane activities of everyday life in chapter one, "The Life of Your Brain." A tour of brain development, brain maturation, and brain aging will follow in chapter two, "Seasons of the Brain." This chapter leads to the central question of the book: What enables the remarkable feats of the mind powered by an aging brain? In chapter three, "Aging and Powerful Minds in History," I will amplify the point by reviewing the lives of several historical personalities remarkable for their pivotal roles in society despite their age and, in some cases, despite their dementia. The resilience of the brain to the effects of age-related decay is greater than most people realize, and you are likely to find some of the examples nothing short of amazing.

We will then proceed to examine the coveted mental attributes of aging—wisdom, expertise, and competence (chapter four, "Wisdom Throughout Civilizations"). Then we will be ready to introduce one of the central concepts of the book—the concept of pattern recognition. We will examine various types of pattern recognition and the role they play in the workings of the human mind. Language is also a pattern-recognition device, but many other such devices operate in human cognition (chapter five, "Pattern Power").

Now it is time to examine how patterns are formed in the brain, and the relationship between patterns and memories (chapter six, "Adventures on Memory Lane"). As it turns out, all patterns are memories but not all memories are patterns. Exactly what distinguishes patterns from other kinds of memories, and what makes patterns less vulnerable than other memories to brain decay, will be the subject of chapter seven, "Memories That Do Not Fade."

How does the well-developed pattern-recognition machinery aid us in everyday life, and what ensures the emergence of such mental machinery? This will be discussed in chapter eight, "Memories, Patterns, and the Machinery of Wisdom." Here we will also introduce a pivotal distinction between "descriptive knowledge" (dealing with the question "What is it?") and "prescriptive knowledge" (dealing with the question "What shall I do?").

Prescriptive, "what shall I do" knowledge is critical to our success in virtually every endeavor. The ability to accumulate and store such knowledge depends on the brain's frontal lobes, which tend to be particularly susceptible to age-related decline. The pivotal role of the frontal lobes in cognition will be the focus of chapter nine, " 'Up-Front' Decision-Making."

Duality is one of the main features of brain design and its enduring enigma. Why is the brain divided into two hemispheres?

Numerous theories and speculations have been offered to account for this fundamental feature of brain design, but none of them has been able to unravel the enigma. We will examine a radically new idea about brain duality: The right hemisphere is the "novelty" hemisphere and the left hemisphere is the repository of well-developed patterns. This means that as we age and accumulate more patterns, a gradual change in the hemispheric "balance of power" takes place: The role of the right hemisphere diminishes and the role of the left hemisphere grows. As we age, we rely increasingly on the left hemisphere; we use it more. This radically new way of understanding the brain's duality throughout the life span will be addressed in chapter ten, "Novelty, Routines, and the Two Sides of the Brain," and in chapter eleven, "Brain Duality in Action."

Division of labor between the two halves of the brain is not limited to cognition. Emotions are also lateralized: Positive emotions are linked to the left hemisphere and negative emotions are linked to the right hemisphere. What does this have to do with different cognitive styles and with aging? This will be the focus of chapter twelve, "Magellan on Prozac."

Aging affects the two halves of the brain differently: The right hemisphere "shrinks" but the left hemisphere shows greater resilience. This is addressed in chapter thirteen, "The Dog Days of Summer." What is behind this mysterious disparity? The answer lies in the lifelong brain plasticity, discussed in chapter fourteen, "Use Your Brain and Get More of It." Contrary to the beliefs held by most scientists until very recently, new nerve cells (neurons) are born in the brain as long as we live. The birth of new neurons and where in the brain they end up are regulated by mental activity. The more we use our brain, the more new neurons we grow, and these new neurons end up in the most-used parts of the brain. As we age, we increasingly use our left hemisphere, which in turn protects it from decay.

This leads to a startling conclusion, deemed fantastic even a few years ago: You can increase your brain longevity by exercising your brain. In chapter fifteen, "Pattern Boosters," we will introduce the various forms that brain exercise may take.

We conclude our exploration in the epilogue, "The Price of Wisdom." Aging, on balance, is not all bad. In fact, it may be something to look forward to and to enjoy. If we value wisdom, then aging is a fair price to pay for it.

So, let's proceed with our exploration of the paradox of wisdom as we age.

1

THE LIFE OF YOUR BRAIN

It's the Brain, Stupid

Most people don't think of wisdom, or for that matter competence or expertise, as biological categories, but they are. Most people understand, generally and vaguely, that our mind is the product of our brain. But it is not always easy to realize just how intimate the relationship is. Despite their acceptance of the mind-brain connection as an abstract proposition, most people don't quite get it on an everyday level. This is a recalcitrant vestige of "mind-body dualism," a philosophical doctrine most closely (even though some students of philosophy say unfairly) associated with the name of René Descartes, according to which the brain and the mind are separate and the mind exists independently of the body. Numerous volumes have been written on the subject, including the excellent books *Descartes' Error*, by Antonio Damasio, and *The Blank Slate*, by Steven Pinker. The centuries-long inability to grasp the idea that the mind *is* the product of the body has inspired the vivid images of homunculus, a little creature sitting inside our brain and doing the hard work of thinking, and of the "Ghost in the Machine." In my earlier book *The Executive Brain* I lamented that even though "today a literate society no longer believes in the Cartesian dualism between body and mind . . . we shed the vestiges of the old misconception in stages" and continue to have difficulties with fully embracing the idea of brain-mind unity when it concerns the highest reaches of our mental life.

I was surprised, even shocked, to discover how fragile and skin-deep this understanding often is. This became starkly apparent a few years ago, when some colleagues and I launched an educational workshop about the brain, titled "The Mind-Brain Institute." The purpose of the workshop was to inform the general public about the basics of brain science, about what may go wrong with the brain and how this may affect the mind, and about the current treatments of various brain disorders. To our great astonishment, the public reaction was often one of incomprehension. "What does the mind have to do with the brain?" was the rhetorical question I heard more than once, to my utter disbelief. Similarly, when in a public lecture about memory I mentioned the brain, a question came from the audience, sounding dismayed more than genuinely inquisitive: "What does memory have to do with the brain?"

Even more incredibly, I encountered similar incomprehension from a much more rarified audience when I was asked to participate in a high-powered symposium on the secrets of extraordinary achievement. The symposium panel was an international "Who's Who" of superachievers: world-renowned scientists, corporate leaders, Olympic champions, famous artists, and high-profile political figures. One by one, these undisputed "champions" in their chosen fields of endeavor were standing up at the podium sharing insights into the secrets of their own achievement. The consensus was rapidly building that the key to achievement is the convergence of two ingredients: Talent in a specific field was unanimously identified as one ingredient of achievement. The presence of certain personality traits, such as drive and the ability to focus on a distant goal, was with equal unanimity identified as the other. The symposium participants agreed that without a special talent there can be no significant achievement and that the special talent is something one is born with, the biological destiny of the few. After all, everyone accepts as a given that hard work alone will not make you a Mozart, a Shakespeare, or an Einstein. But the other ingredients of extraordinary success, drive and ambition, were "up to the individual," the speakers maintained one after another, as if the person in question was a Platonic, extracorporeal entity.

When it was my turn to speak, I tried to convey the idea that "drive" and "the ability to focus on a lofty goal" are also biologically-based attributes, at least in part, and that one of the reasons people vary in these attributes is because their *brains* are different. Personality, I maintained, as I had done in front of various audiences before, is not an extracranial attribute. It is a product of your brain.

My admonition was met with a stone wall of silence, then impatience, and after a few minutes a comment came from a fellow panelist, an illustrious, internationally renowned diplomat: "Professor Goldberg, what you are saying is extremely interesting, but this conference is *about the mind, it is not about the brain.*"

As my jaw dropped in disbelief that such a basically ignorant comment was possible in this highbrow company, I was contemplating a spirited rebuttal in defense of the mind-brain connection, but decided to let it go, for reasons social rather than intellectual.

The simple message I am trying to convey is this: Just as the slightest movement of your body depends on the work of a particular muscle group, so, too, even the most minute, seemingly elusive mental activity calls upon the resources of your brain. And even the simplest of mental activities may be disrupted by brain disease. So as we embark with humility but also with fortitude on our exploration of the seasons of the mind at different life stages, and of the nature of wisdom, we must regard it as a matter of the brain. To borrow from our political folklore, "It's the brain, stupid" is the main theme of this book. Please don't take it personally.

Is the aging of our brains all gloom and no triumphs? I don't think so. In fact, I will use all the mental vigor left in my own aging brain to promote the thesis that the aging of the mind has its own triumphs that only age can bring. That is the central message of this book.

It is time to stop thinking about the aging of our minds and our brains solely in terms of mental losses, and losses alone. The aging of the mind is equally about gains. As we age, we may lose the power of our memory and sustained concentration. But as we grow older, we may gain wisdom, or at least expertise and competence, which is nothing to sneer at either. Both the losses and the gains of aging minds are gradual rather than precipitous. Both are rooted in what happens in our brains. There have been enough books written about the losses of aging minds. This book is devoted to the gains, and the balance between the losses and the gains.

Our culture demands a happy ending to every story. As a product of a harsher environment in my youth, I find this amusing to this day, despite the fact that I have lived this side of the Atlantic for three decades. I recall a television interview I watched after a particularly cataclysmic event of recent years. After a talking-head expert painted a dramatically stark and unfortunately accurate picture of the issue at hand, the interviewer, a famous TV personality, said with a tinge of impatience and even entitlement: "But what can you say to reassure the American public?" At which point I said to myself, *What an interesting cultural idiom! Give me a happy ending or else!*

Reassurance is not always a good thing. There are circumstances when grabbing the public by the scruff of its collective neck, so to speak, and shaking it up with alarm will do more good in the long run. But on the issue of aging the public has already received its therapeutic shake-up dose. We hear constantly about the scourges of dementia and Alzheimer's disease, and about the symptoms of neuroerosion,[1] the encroachment of forgetfulness and increasing mental fatigue. Unfortunately, these scourges are real. But it is time to look for good news, providing that the good news is also real and not a phony "reassurance" ploy.

Explaining Wisdom

Wisdom is the good news. Wisdom has been associated with advanced age in the popular lore of all societies and through history. Wisdom is the precious gift of aging. But can wisdom withstand the assault of neuroerosion, and for how long?

This raises a question about the nature of wisdom. In our culture we use the word frequently and reverently. But has wisdom ever been sufficiently defined? Its neural basis understood? *Can* the phenomenon of wisdom be understood in principle in biological and neurological terms, or is it too elusive and multifaceted to be tackled with any degree of scientific precision?

Without claiming any particular wisdom of my own, I believe I can contribute to this understanding by enlarging on my earlier introspections, which help elucidate the nature of wisdom,

---

[1] I have coined the term *neuroerosion*, and by extension *neuroerosive*, to fill a certain perceived gap. It is common to refer to certain disorders eventually resulting in dementias as *neurodegenerative*. But this term is both too narrow and too ominous. It implies a very specific set of disorders characterized by primary neuronal atrophy. The terms *cerebrovascular* or *multiinfarct*, often used to refer to certain other disorders eventually resulting in dementias, also have very specific, narrow connotations, implying primary disease of the blood vessels in the brain. Neuroerosive is intended to be a generic term, covering all the specific possibilities and devoid at the same time of the sense of finality attached to those other terms. It is similar in its scope and implications to the term Mild Cognitive Impairment (MCI), which has become increasingly popular lately, without the latter term's antiseptic, clinical ring, as well as the risk of a trademark infringement lawsuit from MCI, the troubled telecommunications company.

or at least one important aspect thereof. The train of thought and the argument developed in this book will flow from this introspection and this insight.

With age, the number of real-life cognitive tasks requiring a painfully effortful, deliberate creation of new mental constructs seems to be diminishing. Instead, problem-solving (in the broadest sense) takes increasingly the form of pattern recognition. This means that with age we accumulate an increasing number of cognitive templates. Consequently, a growing number of future cognitive challenges is increasingly likely to be relatively readily covered by a preexisting template, or will require only a slight modification of a previously formed mental template. Increasingly, decision-making takes the form of pattern recognition rather than of problem-solving. As the work by Herbert Simon and others has shown, pattern recognition is the most powerful mechanism of successful cognition.

Evolution has resulted in a multilayered brain design, consisting of old subcortical structures and a relatively young cortex with a particularly young subdivision appropriately called the neocortex. The cortex of the brain is in turn divided into two hemispheres: right and left. The passage from problem-solving to pattern recognition changes the way these different parts of the brain contribute to the process. Firstly, cognition becomes more exclusively neocortical in nature and increasingly independent of subcortical machinery and of the machinery contained in the old cortex. Secondly, the balance of our use of the two hemispheres of the brain shifts. As I will show, in neural terms this probably means a decreasing reliance on the right hemisphere of the brain and an increasing reliance on the left cerebral hemisphere.

In neuroscientific literature, the cognitive templates that enable us to engage in pattern recognition are often called *attractors*. An attractor is a concise constellation of neurons (nerve cells critical for processing information in the brain) with strong connections among them. A unique property of an attractor is that a very broad range of inputs will activate the

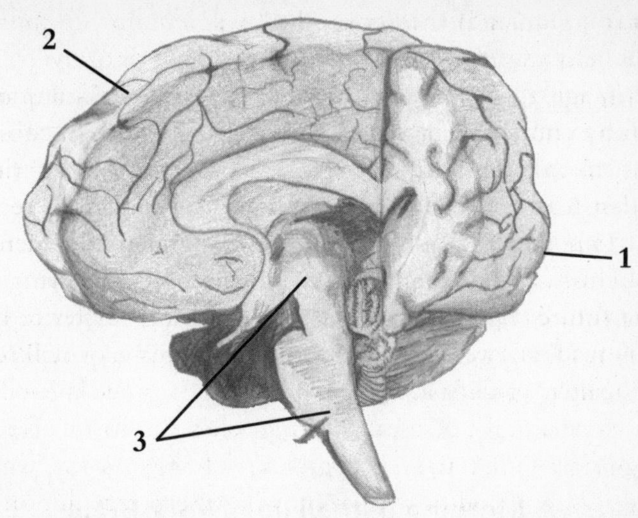

FIGURE 1. **Human Brain.** *Cerebral hemispheres (1 and 2) and subcortical structures (3). The frontal portion of the left hemisphere is removed, exposing the brain stem and the diencephalon.*

same neural constellation, the attractor, automatically and easily. In a nutshell, this is the mechanism of pattern recognition.

I believe that those of us who have been able to form a large number of such cognitive templates, each capturing the essence of a large number of pertinent experiences, have acquired "wisdom," or at least a certain crucial ingredient thereof. (As I write this, I hear the indignant howling of critics from various corners of science, humanities, and social activism, accusing me of scandalously gross oversimplification, so I am hedging my bets).

By the very nature of the neural processes involved, "wisdom" (at least in my admittedly narrow definition of it) pays dividends in old age by allowing relatively effortless decision-making requiring only modest neural resources. That is, modest as long as the templates have been preserved as neural entities. Up to a point, wisdom and its kin qualities, competence and expertise, may be impermeable to neuroerosion. These will be the main themes of the book.

But before we delve into the brain mechanisms of the cognitive gains in aging, we need to dispense with several preliminaries. We need to examine the nature of wisdom as a psychological and social phenomenon. We need to establish to our satisfaction, whether it is truly the case that a powerful mind may persevere and, to a point, prevail and triumph, even in the face of neuroerosion. This will be the book's humanistic foundation and its point of departure, followed by a journey into the mysteries of the neural machinery of wisdom, competence, and expertise, and of the cognitive gains in aging.

A Morning in the Life of Your Brain

Before we delve into these intriguing issues, let's have an introduction to our own brain. How does this magnificent piece of biological hardware work, and how do you use it in your everyday activities? Let's start from the beginning, so to speak, and consider a morning in the life of the brain.

The alarm has just rung, rudely assaulting your *brain stem,* your *thalamus*, and your *auditory cortex*. The sound awakened you out of deep sleep, which means that the auditory signal somehow activated a particular part of the brain stem, the *reticular formation* in charge of general arousal. Had it been a different sound—a dog barking, a fire-engine siren blaring, raindrops falling—you would have sighed with annoyance and gone back to sleep. But reluctantly, you open your eyes. Your auditory cortex, with the help of certain *thalamic nuclei*, has recognized the sound for its source: it is an alarm clock. And your frontal lobes, the superego of the brain, tell you that this is important and you must get up.

You get out of bed and look out the window. You are barely awake, but your *visual cortex* is already working away allowing you to appreciate the beautiful morning outside. Don't take it for

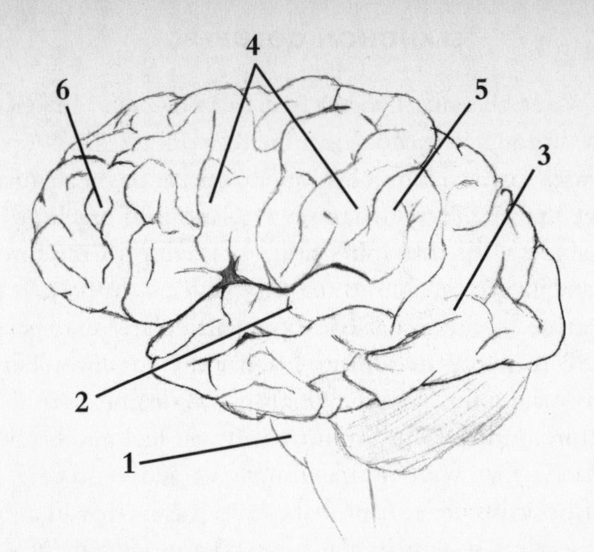

FIGURE 2. **Different Brain Regions: What They Do.** *Waking up (1); recognizing alarm clock (2); spotting the toothbrush (3); using it (4); checking time (5); planning the day ahead (6).*

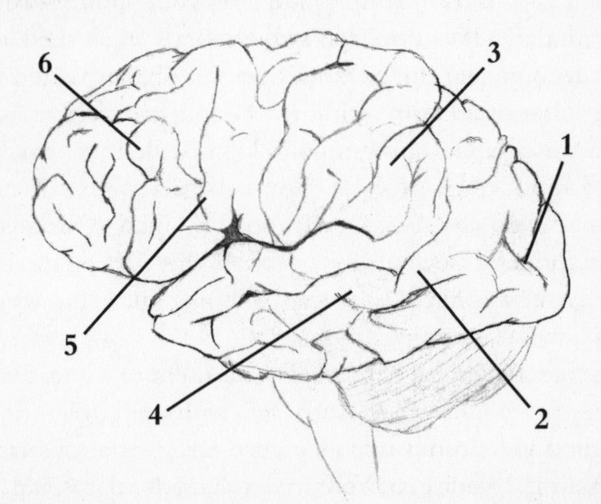

FIGURE 3. **Different Brain Regions: What Happens When They Are Damaged.** *Anton's syndrome—cortical blindness (1); visual object agnosia—inability to recognize common objects (2); ideational apraxia—loss of skilled movements (3); Wernicke's aphasia—affects mostly object words (4); Broca's aphasia—affects mostly action words (5); executive deficit—impaired planning (6).*

granted. When the visual cortex is damaged, *cortical blindness* develops even though the eyes continue to work just fine. A patient afflicted with cortical blindness (due to stroke or mechanical injury to the brain) will be able to see gradations of brightness, will even be able to tell that something is moving in the environment, but will not be able to identify objects. In certain cases, when damage to the visual cortex is particularly extensive, the patient will even lose the ability to realize that his vision has been lost. This condition is known as Anton's syndrome.

It is sunny outside your window and you feel good. "Feeling good" means that your *left frontal lobe* is active, since it is in charge of positive affect. It probably also means that in a particular biochemical system in the brain, the *neurotransmitter dopamine* is kicking in.

As you walk into the bathroom, you survey the familiar objects: your toothbrush, your toothpaste, your mouthwash, your razor. Familiar? Of course, you know exactly what these objects are. But recognizing things as meaningful objects would not be possible without a brain region in the left hemisphere, roughly between the occipital and temporal lobes, called the *visual association cortex*. This part of your brain is hard at work, despite the fact that you go about your bathroom business effortlessly and casually, maybe not even fully awake. If this part of the brain is damaged, you would continue to see things, but fail to recognize them as familiar, meaningful objects.

This is precisely what happened to a patient of mine, a middle-aged woman who walked into the bathroom one morning, looked around, and did not recognize any of the objects lying about. Alarmed, she had herself driven to the local hospital, where a CT scan was immediately performed. It turned out that she had had a stroke the night before, which affected her visual (occipital) cortex, causing a condition called *visual object agnosia*. It can also be caused by head injury or dementia. To help restore the function of her brain, a comprehensive program of cognitive rehabilitation was indicated, which is how she became my patient.

Luckily, your visual association cortex is doing just fine. You are reaching for the brush with your hand. The odds are about nine to one that it will be your right hand because approximately 90 percent of the population is right-handed. The *motor cortex* in your left hemisphere (the pathways between the brain and the body are mostly crossed) rushes into action, and so do your *cerebellum* and your *basal ganglia*. Without these brain structures, even the simplest, most automatic and effortless movement would be impossible.

You grasp the toothbrush in your hand—it seems like a simple activity, despite all this neural commotion—and lo and behold, you did it right: you took the toothbrush by its handle and not by the brush itself. But to accomplish this ridiculously simple feat, complex neural machinery had to kick in. It is not enough to know what the object is, one must also know how to use it. The knowledge of the motor program corresponding to the use of common objects is stored in the *parietal lobe*, mostly in the left hemisphere. Damage to this part of the brain due to stroke or Alzheimer's disease often leads to *ideational apraxia*. The patient loses the ability to use common objects according to their function and instead begins to manipulate them randomly, like a newcomer from a different culture, where the object does not exist and therefore cannot be meaningfully recognized. Sometimes this deficit takes the peculiar form of *dressing apraxia,* when the patient loses the ability to put on his or her clothes correctly. This, too, is commonly seen in dementias.

But your neural machinery is in top shape, and after finishing in the bathroom you have your business suit on in no time at all. Outside the city is coming to life and loud music begins to blare from a nearby construction site and in through the kitchen window. "What trash," grumbles your *right temporal lobe*, in charge of processing music, causing you to wince. Strictly speaking, the right temporal lobe generates the aesthetic judgment, but it is your left hemisphere that parlays it into words.

Time for a hasty cup of coffee and the morning newspaper.

As you scan the front page, your left hemisphere is abuzz. The *left temporal lobe* is processing and understanding nouns, the *left frontal lobe* is processing and understanding verbs, and the *left parietal lobe* is processing grammar. Damage to these parts of the brain causes various forms of *aphasia*. Meanwhile, the *prefrontal cortex* is figuring out frenetically what the news of an impending recession means for your own job. The NASDAQ is down the third day in a row and so is the Dow Jones Industrial Average. You can remember what the newspapers said a few days ago when the markets were still up, which means that, unlike your stock portfolio, your *hippocampi* are still OK. The hippocampi, of course, are critical for learning new information.

Despite the sunny spring morning, the stock exchange set you temporarily in a foul, seething sort of mood, and your *amygdala,* in charge of emotions, briefly lights up. For reasons to be explained later, it is likely to be your *right amygdala*.

As you are rushing out the door, you are figuring out feverishly how to juggle five meetings and three conference calls, all scheduled for today. Your *prefrontal cortex*, responsible for organizing things in time, is hard at work, trying to do the near-impossible: to sequence eight activities with clockwork precision and no room for slack.

In the elevator you notice an unfamiliar face. A new tenant in the building? It was your *right hemisphere* that analyzed the face in the elevator and concluded that it was a new one.

You catch a cab and look at your watch. Your *parietal lobe* quickly takes in the dial display. You are going to make it to your office on time, more or less. But as you are about to sigh with relief, you notice that the cab driver has just taken a wrong turn. No wonder, you think, he is probably just off the boat and does not know the city. You quickly take control of the situation and attempt to guide the driver back on the right track. That takes a coordinated action of the *frontal lobe* (sequencing) and the *parietal lobe* (spatial information). But the good man does not understand what you are saying, since he does not speak English! You improvise by using universal sign language to direct him (your *frontal, parietal,* and *temporal* lobes are working furiously in concert).

You are finally there. You quickly pay the driver and count your change (*left parieto-temporal* part of the brain, which if damaged produces a deficit called *acalculia*, loss of computational skills). You made it. Your brain can relax for a few precious moments while you are waiting for the elevator.

So what is going on here? Your working day has not even started yet, and your brain has already been hard at work. The few trivial, effortless, routine morning activities required the involvement of virtually every part of the brain. And I will be the first to admit that my account of the morning in the life of the brain was a gross oversimplification, highlighting just a few main actors on the stage of the brain theater, just a few main players in the cerebral orchestra. In reality, every stage of my account involved a myriad of supporting actors in addition to the lead ones, all blending into complex and intricate cerebral ensembles, different at every moment of our lives, and fluidly communicating with one another in time.

In scientific terms, these ensembles are called *functional systems*, a term introduced by the great Jewish-Russian neuropsychologist Aleksandr Romanovich Luria (more about him later). Even though neuroscientists had inferred the existence of such intricate, dynamic processes long ago, it has become possible to actually observe them only recently, with the advent of powerful new technologies of functional neuroimaging, which literally offer us a window into the inner workings of the living, acting, thinking brain.

Just Watching TV

To flesh out the notion of a functional system, many aspects of the mind and thus many parts of the brain working together in concert, let us consider the following, so-familiar situation: just watching TV.

It is the end of a late Saturday afternoon, and you are sitting in your living room basically not doing much at all. Your dog is snoozing at your feet. You are nursing your cup of coffee, or whatever your favorite late-Saturday-afternoon drink may be. You are doing nothing, really, just watching CNN.

In the midst of this blissful nothingness, your brain is hard at work, engaged in a complex and fluid ensemble of activities, while you are ostensibly lazing around. Your visual and auditory cortex are abuzz, processing the images on the screen and the voice of Christiane Amanpour delivering the breaking news of the day. For simple signal detection, older subcortical structures in the brain stem and the thalamus may suffice, without particularly engaging the neocortex. But this is highly meaningful information and the neocortex is involved.

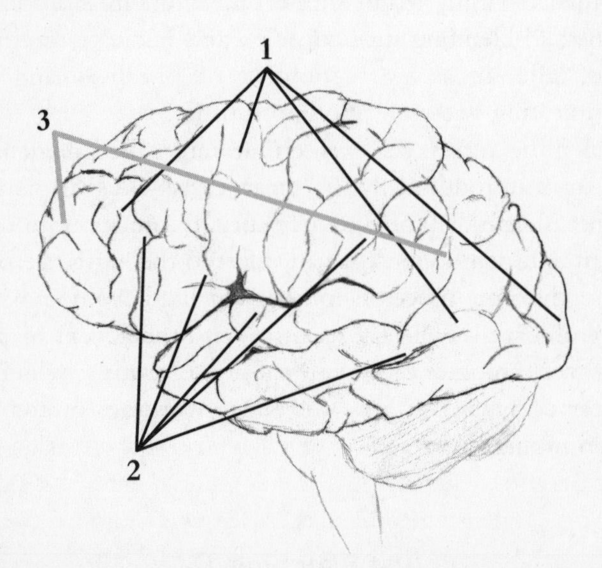

FIGURE 4. **Brain Regions Involved in Watching TV.** *How functional systems work. Examining visual images (1); understanding what the commentator says (2); putting it all together (3).*

Indeed, digesting the news about a tense confrontation half the world away takes up the resources of much of the brain. The verbal content of Amanpour's narrative engages much of your left hemisphere. (This assumes you are right-handed, and if you are left-handed, the odds are still approximately six to four that your left hemisphere is mostly in charge of language). First engaging the part of the temporal lobe called the *superior temporal gyrus* in charge of speech sound perception, it then engages much of the rest of the left hemisphere.

Language is a cultural tool of incredible complexity and versatility. We often think of language as a means of communication. It is certainly that, but also much more. As we will discuss later, language is a means of conceptualization, of information compression, which enables us to represent complex information in compact codes. The brain machinery of language is highly distributed. As already mentioned, the meaning of object words (nouns) is stored in the left temporal lobe close to the visual cortex. That makes sense: Our mental representations of objects are based mostly on vision. The meaning of action words (verbs) is stored in the left frontal lobe close to the motor cortex. That also makes sense: Our mental representations of skilled movements involve those parts of the brain. Complex statements establishing relations between things are processed in the part of the left hemisphere where the temporal and parietal lobes come together—the *left angular gyrus.*

Damage to these different parts of the brain will impair language in different ways, to use technical language, will produce different forms of *aphasia*, depending on where exactly in the left hemisphere it occurs. The causes of such damage vary: It could be stroke, head injury, or dementia. Indeed a particular form of language disorder, called *anomia* (loss of the use of words), is among the early symptoms of Alzheimer's disease.

But the right hemisphere is not left out of the action, either. As Christiane Amanpour's voice rises to an urgent crescendo, it is the *right hemisphere* that detects the feeling of alarm conveyed by it. While the left hemisphere is in charge of most aspects of language in an adult brain, the right hemisphere is in charge of *prosody.* Prosody is information conveyed through verbal communication, but by means of intonation and inflection rather than by the literal meaning of words. It is what we call the "emotional tone." (Dysfunction of the right hemisphere, as in Asperger's syndrome, impairs the ability to process such "extralinguistic" contextual information. As a result, the patient's behavior becomes mechanical, awkward, and often inappropriate, devoid of subtlety and fluidity.)

Your dog has also sensed the urgency in the commentator's voice (I don't know with which hemisphere of his brain; hemispheric specialization has not been extensively studied in animals, although I have advocated such research for years) and began to growl. You recognize his canine growl, as opposed to any other sound in the environment, without taking your eyes off the TV screen. This was also accomplished through the left hemisphere, the *left temporal lobe* to be precise. Damage to the left temporal lobe produces not only aphasia, but also an inability to identify environmental sounds by their sources. This often overlooked condition is called *auditory associative agnosia.*

Meanwhile, the *visual cortex* has been busy all along taking in the images on the television screen. Since you are in excellent neurological health, you easily take in information both from the left and the right half of the screen. You can do this because both hemispheres of your brain are working just fine, and the connection between them, a thick bundle of pathways called the *corpus callosum* (the latter word from Latin for callus), is intact. Damage to one hemisphere, particularly to the parietal lobe, often produces *visual hemiinattention* or even outright *visual hemineglect.* A patient afflicted with visual hemiinattention has difficulty attending to the information appearing in one half of the visual field—the half *opposite* to the side of brain damage. Visual hemineglect is even more severe than visual hemiinattention, one half of the visual field being completely ignored. Left visual hemiinattention or hemineglect (caused by damage to the right hemisphere) is usually much more severe than right visual hemiinattention or hemineglect (caused by damage to the left hemisphere).

What's even more interesting is that the patient is often unaware of left hemineglect or left hemiinattention. Such unawareness of deficit is itself a neurological symptom, usually caused by damage to the right hemisphere, and it is called *anosognosia.* Anosognosia is a source of all kinds of hazards, since the patient may be unaware of *any* deficit, not just hemineglect or hemiinattention. Imagine a driver afflicted with visual hemiinattention yet unaware of it. Unfortunately, this is not that uncommon in patients who have suffered a stroke in the right hemisphere. Despite the fact that it will be obvious to everybody else around, any attempt to convince the patient of the impairment will likely meet with failure. This is often called "denial," but strictly speaking it is not. "Denial" implies an intact ability to know and a choice not to know. In anosognosia the capacity for knowing one's own deficit is genuinely lacking due to brain damage. A patient will often insist on driving and carrying on with other activities that imperil himself and other people.

In a highly protective environment, the effects of hemineglect or hemiinattention may be more comical than tragic. I will never forget an elderly man in a nursing home who suffered a right-hemispheric stroke with left hemineglect and ranted indignantly about the conspiracy of nurses. He was furious that his fellow patient sitting across from him at the cafeteria table was getting a steak, while all he was getting was mashed potato—an outrageous inequity indeed. The key to this apparent injustice was simple. The kitchen personnel had the habit of placing the steak on the left side of the tray and the mashed potato on the right side of the tray. So the old gentleman always saw the potato on right side of his tray and the steak on left side of the tray in front of the fellow sitting opposite him. And it was impossible to get the old man to comprehend that the problem was within and not outside, until the nurses learned to flip the tray in front of him. The patient remained convinced that he was a victim of dirty tricks and that nothing was wrong with *him*. Yet aside from his dinnertime ire, he was the happiest, go-luckiest patient on the unit.

Unlike the old man, your visual fields are in good order, left, right, and center. So you are able to scan the whole television screen and follow the important details. The ability to scan a detail-rich visual scene extracting important information from wherever it may appear in the environment is ensured by a region of the frontal lobes called the *frontal eye-fields.* They are firing away as you are relating Christiane Amanpour's commentary to the images on the screen.

As you do so, you are processing the specific visual images on the screen. You recognize them as representations of meaningful objects: houses, cars, trees . . . and unfortunately tanks, guns, and such. This throws into action another part of the visual cortex, your visual association cortex, mostly in the left hemisphere, as already mentioned.

You also see faces—smiling faces, anxious faces, happy faces, angry faces, faces of unknown people in a faraway country. As you stare at them, trying to have a glimpse into the minds behind the faces, the temporal lobe of your right hemisphere is hard at work. This part of the brain has been shown to be in charge of facial recognition.

But curiously, the face of Christiane Amanpour is processed mostly by your left hemisphere. A peculiar division of labor takes place in the brain. The right hemisphere is better at dealing with novel, unfamiliar information, and the left hemisphere is better at dealing with familiar information. This is true for most kinds of information, so that the faces of strangers are processed on the right and the faces of public figures, or family members and friends whom you encounter all the time, are processed on the left.

With the "Breaking News" report in progress, a map appears in the upper right corner of the TV screen to highlight the place of the events. This brings into action your *spatial, parietal lobe* on its junction with the *visual, occipital lobe.* Neuroscientists distinguish between the "what" and the "where" visual systems in the brain. The "what" system, on the junction of the occipital and temporal lobes, is in charge of object recognition. The "where" system, on the junction of the occipital and parietal lobes, is in charge of location information.

As the visual images and the reporter's narrative blend seamlessly into a story, you are not even cognizant of which information comes in through the eye and which information comes through the ear. It all becomes intertwined and interwoven in your mind. This is because your heteromodal association cortex is doing its job properly and efficiently. This part of the brain is in charge of putting together streams of information coming in through different senses and integrating them into one neural multimedia theater. Among the most recent to develop in evolution, this part of the brain is particularly vulnerable to Alzheimer's disease and other dementias.

This is the third time the region is in the news this week, you are saying to yourself as you follow the Breaking News. In order to reach this conclusion, you must be able to relate the current events as presented in the news today with your memories of the news over the past few days. You have just successfully employed your recent memory, for which the hippocampi are particularly important. Hippocampi are also particularly vulnerable in Alzheimer's disease. In fact, Mony de Leon and his colleagues at the Aging and Dementia Research Center of New York University's School of Medicine have developed innovative techniques using fine measurements of the hippocampal size based on magnetic resonance imaging (MRI) as an early predictor of vulnerability to Alzheimer's disease.

The good news brought to us by state-of-the-art neuroscience research is that new neurons tend to develop in the hippocampi. What's particularly exciting is that the rate at which the new neurons appear in the hippocampi can be influenced by cognitive activities and by exercising your brain. We'll explore this in later chapters.

As the news is being delivered, you are trying to figure out what will happen in the conflict-ridden region next. A game of prediction, like a game of chess, is a tricky business. You need to assess the overall context and to put yourself in the place of each of the main players. You need to plausibly surmise what *they* think of the situation. Napoleon understood this very well, when he admonished his marshals: In anticipating the adversary's move, don't expect him to do what *you* consider to be his optimal move. Try to figure out what *he* considers his optimal move from his own perspective, given his own history, and with the information likely to be available to *him,* not to you. The ability to put yourself in someone else's "mental shoes" is called by cognitive neuroscientists the capacity to form *the theory of mind.*

These complex abilities—to plan, to anticipate, to form the theory of mind—are all very young in evolutionary terms. They are present only in humans in a developed form, and one might say, they are what makes us human. All these complex functions, which we have begun to understand only recently, are controlled by the prefrontal cortex. I wrote about it extensively in my earlier book, *The Executive Brain.* The youngest and most complex part of the human brain, it is also the last to develop. It is fully developed only by the age of eighteen or possibly even as late as thirty. This validates the custom shared by most modern cultures, according to which the age of eighteen or thereabout is the age of legal maturity, and the eligibility for highest elective offices requires an even more advanced age. The prefrontal cortex is very vulnerable in a broad range of neurological and psychiatric disorders, such as dementia, schizophrenia, or traumatic brain injury. Dysfunction of the prefrontal cortex has also been implicated in such less devastating but nonetheless disruptive conditions as Attention-Deficit/Hyperactivity Disorder and Tourette's syndrome.

Your own prefrontal cortex was nudged out of its slumber the moment you began to play the game of crystal ball, trying to make political predictions. And so was your *anterior cingulate cortex,* a brain structure closely linked to the prefrontal cortex, which is particularly active in situations of uncertainty.

But you know your limitations and can spend only so much time playing the game of crystal ball, a game that even Napoleon eventually lost. Your attention is drifting and you are beginning to feel sleepy. That means that your *ascending activating reticular formation,* a very important structure in charge of keeping the brain aroused and alert, has had it for now.

You yawn, stretch, and turn off the TV set. The thought of taking your dog for a walk crosses your mind, but then you decide to stick around and refill your drink. Your hypothalamus, amygdala, and *orbitofrontal cortex* have all lit up—the mechanisms of basic gratification. . . . Life is that simple on a Saturday afternoon.

2

SEASONS OF THE BRAIN

What Happens to the Brain Happens to the Mind

Now that we are done with this casual review of your brain in action, stand back and think (your brain again). If activities as trivial as a day-in, day-out morning routine or watching the news on television are so demanding of brain resources, can you imagine the brain machinery behind the complex professional activities of a physician or an engineer, the intellectual rigor of a mathematician or a chess player, or the creative surge of a violinist or a dancer? Cognitive neuroscience is only beginning to address these issues, but it is no longer possible to think or talk about the mind without the brain, or about the brain without the mind.

As a typical reader of this book, you are not a brain scientist, but you are a brain user, a consumer of brainpower, so to speak. And the odds are that you have not been particularly inquisitive about the inner workings of your brain. This is a curious phenomenon, and it concerns all of the human body, not just the brain. Ironically, most of us generally do not care about our body, as long as it leaves us alone, does not ache, hurt, itch, or malfunction, and allows us to feel good. If Johnny contracts hepatitis A from bad oysters, he does not go to the doctor because his liver enzymes are elevated and viral titers are up; he goes because he feels lousy and tired, and because his face and eyeballs have turned yellow—not a highly valued trait on the dating circuit.

Even though Johnny does not particularly care to know about the inner workings of his body, he accepts the general premise that how he feels depends on, among other things, the condition of his liver, which has to be dealt with in order for Johnny to feel good again and regain a desirable complexion. But when it comes to the mind-brain relationship, the closeness of this link does not seem to have trickled into the public awareness yet. The general public is only beginning to appreciate the fact that any assault on the brain will affect your mind.

But is the inverse true? Can we *improve* the quality of the mind by improving the function of the brain? If the answer to this question is "yes," then Johnny should start learning how to take care of his brain, just as, in the last few decades, he has embraced the notions of healthy physical living (raw oysters notwithstanding). In this book, I will argue that what happens to one's brain as one ages depends to a great extent on what one does with it at a younger age. I will also argue that it may be possible to improve one's mind by improving one's brain even at an advanced age. I will discuss how this happens in everyday life and what can be done to accomplish it better in a more structured manner.

First, though, we need to understand the natural processes in the brain throughout the life span. "Seasons of the mind" or *seasons of the brain* is, of course, a metaphor, but not too far-fetched a metaphor. The brain and the mind go through stages in the course of a lifetime. Like the seasons of the year, the seasons of the mind are not separated by clear-cut absolute boundaries, but morph gradually and seamlessly into one another. So any attempt to link these boundaries to precise chronology is a matter of convention rather than of real biological discontinuities. Just as the change between seasons may vary from year to year (early summer one year, late spring another year), so too the exact timing of transition from one "season of the mind" to the next varies somewhat from person to person. To complicate matters even further, not all aspects of the mind and the brain move through the stages in perfect synchrony. This means that how exactly you set the boundaries between the stages depends to a large degree on your choice of the criteria. Unlike the four seasons of the year, it is common to speak about three seasons of the brain: *development, maturity*, and *aging.*

Developing Brain

The first season, *the season of development,* is when the main cognitive abilities and skills are formed, which is characterized by dramatic changes in the brain. This season begins before we are born and extends into the third decade of our life. Brain development is a complex and multifaceted process. It starts with *neurogenesis*, the birth of neurons, which are the brain cells most directly involved in information processing, and their migration, finding their proper places in the complex organization of the brain. For the most part, neurogenesis occurs during gestation, at somewhat different times for different brain structures. It was thought until recently that neurogenesis ran its course and ground to a complete halt sometime during gestation and the first few years of life. By that time, most brain structures have acquired their recognizable shape. Today we know, however, that neurogenesis continues throughout the lifetime, albeit not as vigorously as during the early period.

As the neurons are born and migrate to their proper locations in the brain, connections between neurons begin to develop. These connections, formed as protrusions emanating from the neuron bodies, are called *axons* and *dendrites.* They begin to develop during gestation, and the dendrites begin to sprout through the process called *arborization*. This process culminates during the first years of life.

Synapses, the tiny interfaces between the dendrites and axons emanating from different neurons, are critical for communication between neurons. Their formation is called *synaptogenesis* and its time course varies considerably for different parts of the brain. In the visual cortex, for instance, most of synaptogenesis is complete by the end of the first few years of life. By contrast, the synaptogenesis of the prefrontal cortex extends well into late adolescence and early adulthood.

Production of neural structures is complemented by the elimination of excessive neurons, dendrites, and synapses. This process, known as *pruning* or *apoptosis*, occurs after birth and also unfolds at different time courses for different parts of the brain, the frontal cortex being the last. Pruning is akin to sculpting, a process that the great sculptor Auguste Rodin described as "eliminating everything that does not belong." Pruning is not random, but rather is a consequence of reinforcing heavily used neural structures and letting go of those underused or not used at all. These competitive processes of the brain molding itself are somewhat akin to natural selection, which was captured in the term "neural Darwinism," coined by Gerald Edelman.

Neurons are not the only type of cells found in the brain. In fact, they account for only about one-third of all the brain cells. The remaining two-thirds are taken up by the *glial cells*, which serve various supporting functions and come in two kinds: *astrocytes* and *oligodendrocytes.* At a certain point in development the process of *myelination* begins: Oligodendrocytes begin to wrap around long axons, forming a fatty protective coating called *myelin.* Myelin is white, which gave rise to the term *white matter* (composed of all the long pathways covered with myelin), as opposed to *gray matter* (composed of all the neurons and short local nonmyelinated pathways). Myelin facilitates signal transmission along the axon, greatly enhancing and improving transmission of information within large coordinated neuronal ensembles. Dramatic increase in brain weight during the first years of life is largely due to myelination. The brain structures are not fully functional until the axons connecting them are insulated with myelin, and the time course of myelination varies vastly from structure to structure. As you can probably guess by now, myelination takes the longest in the frontal cortex, extending well into late adolescence and young adulthood, possibly until the age of thirty. The volume of the frontal lobe, and particularly of the prefrontal cortex, continues to grow at least until the age of eighteen and possibly longer, and this growth reflects an ongoing increase in white matter.

If nothing else, this brief review shows that brain development is the interplay among numerous processes unfolding at different time scales. This is a time of great flux in the life of the brain. This is also a time of great flux in the life of the mind—the time of learning, of accumulating the basic fund of mental skills and knowledge, and ultimately the time of forming our identities.

You may have noticed that the frontal lobes, the prefrontal cortex in particular, are the last to complete their biological maturation—only by young adulthood, sometime in the very end of the second decade and possibly even in the third decade of life. Modern society operates on the basis of certain tacit or explicit assumptions about the age of social maturity. This is the age of emergence of the cognitive and personality traits that we associate with social maturity, such as the capacity for impulse control, foresight, and critical self-appraisal. Like the biological maturation of the frontal lobes, these "adult" traits reach their full functionality sometime in the end of the second and the beginning of the third decades of life. Unsurprisingly, this age has been codified in virtually every modern society as the age of transition from social immaturity to social maturity. This is the approximate age (plus or minus a few years) when you are ready to assume a whole range of "mature" rights and responsibilities, such as driving, voting, getting married, buying alcohol, serving in the military, and finally being treated by the legal system as an adult and not a minor. What most people don't realize is that the emergence of these "adult" traits is most likely caused by the maturation of the frontal lobes, a belief shared by an increasing number of neuroscientists. Thus, many neuroscientists find it useful to think of the completion of the maturation of the frontal lobes, particularly the myelination, as the watershed between the first and the second seasons of the brain: the stage of development and the stage of maturity.

Mature Brain

The second season, *the season of maturity*, is characterized by less neural flux and by greater stability of brain structures. This is the age of productive activity, when the emphasis gradually shifts from learning about the world to contributing to and molding the world around us through our individual professional and vocational activities. This is the most extensively studied season of the mind and of the brain. In fact, until a few decades ago our knowledge was limited to this stage. The standard texts of neuroanatomy, neurology, or neuropsychology, as well as dozens of books written for the general public, are mostly about this stage, so there is no point in restating much of this normative knowledge here. Suffice it to say, in our zeal for generalizations we have been treating the mature brain in rather generic terms. This is undoubtedly a useful enterprise, and a reasonable point of departure for any scientific inquiry, but only to a point. While perusing any standard text, you are not likely to encounter any reference to the gender differences in brain organization, let alone to the individual differences. But such differences do exist and we are only now beginning to understand them. From the aerial view of all humanity represented by a composite, we are gradually moving to the understanding of the neural foundations of individuality.

Aging Brain

Then comes the third season, *the season of aging*. What happens to the magnificent brain machinery as we move further through life? How golden is the "golden age"? Oddly, scientists did not endeavor to address this question until relatively recently. Hippocrates himself omitted the brain from the litany of old-age woes in his *Aphorisms.* About this, the leading neuroscientist of aging Naftali Raz observed:

> . . . So overwhelming are the transformations of the aging body and so pervasive are the changes in its basic functions that it may not be particularly surprising that the most famous of the ancient servants of Aesculapius did not find the brain and higher cognitive functions sufficiently important to be included in his list of geriatric troubles.

But the brain *is* affected in aging, even in successful, healthy aging. It would be strange if it weren't because, like every other organ, the brain is of the flesh. Extensive research has been conducted over the last few decades to understand such changes, and today we have a relatively comprehensive picture of what happens to the aging brain, even when the process is unencumbered by neurological illness or dementia. Much of the discussion that follows in this chapter is based on Naftali Raz's own research and on his cogent reviews of the state of affairs in the field of brain aging research.

Some of the changes that occur as the brain ages are global. Both the weight of the brain and its volume shrink by about 2 percent every decade of adult life. The ventricles (cavities deep inside the brain containing cerebrospinal fluid) increase in size.

The sulci (spaces between the walnut-like convolutions of the cortical mantle) become more prominent. All these changes suggest a modest amount of atrophy or shrinkage of brain tissue, even as part of normal aging. The connections between neurons become increasingly sparse (a process known as "debranching"), and so does the density of synapses (the sites of chemical signal transmission between neurons). Blood flow to the brain becomes less abundant, and the oxygen supply to the brain less generous.

Both gray matter and white matter are affected in aging. In the white matter, small focal lesions appear. They are sometimes called *hyperintensities* in the technical parlance of MRI radiology. In most cases, the aging-associated "hyperintensities" reflect vascular illness, but they may also reflect demyelination of pathways. They tend to accumulate with age. The relationship between these focal white-matter lesions and cognitive decline is not a simple linear one, but rather of a threshold nature. Up to a point, they remain benign, but once their total volume reaches a certain level, cognition begins to deteriorate. Some scientists believe that the white matter is more susceptible to the effects of aging than the gray matter.

Against the background of such global changes, certain parts of the brain fare better than others. A number of cortical and subcortical structures are affected, but to different degrees. In the neocortex, the classic neurological rule of "evolution and dissolution," first introduced by John Hughlings Jackson, seems to operate: The phylogenetically (evolutionarily) youngest cortical subdivisions (which develop only at the later stages of "evolution"), the so-called heteromodal association cortex, are affected to the greatest extent by "dissolution" due to aging. These include inferotemporal, inferoparietal, and particularly the phylogenetically most recent prefrontal cortex. By contrast, the phylogenetically older cortical subdivisions, which include the areas involved in receiving raw sensory information and the motor cortex, are least affected. The prefrontal cortex, a subdivision of the frontal lobe in charge of complex planning and the organization of complex behaviors in time, is affected to the greatest extent by aging.

A similar relationship exists between ontogenetic (occurring during a lifetime) development and decay: The brain structures last to develop at the organism's growth stages are the first to succumb to decline with age. In assessing the relative vulnerability of various brain structures, the fate of the pathways projecting from and to these structures is particularly instructive. Therefore, the chronology of pathway myelination is a useful marker both of development and of decline. By this light, the longer it takes for the pathways to myelinate, the more susceptible is the corresponding structure to the effects of aging. Again, the prefrontal cortex comes out as the most vulnerable, particularly its *dorsolateral* subdivision. The changes in the frontal lobes involve the deterioration of both the gray matter and the white matter, as well as the depletion of major neurotransmitters (chemicals in charge of signal transmission between neurons): dopamine, norepinephrine, and serotonin. As was the case in development, the fate of the frontal lobes serves as the watershed between the second and the third seasons of the brain, the stage of maturity and the stages of aging.

Outside the neocortex, the hippocampus and the amygdala are only moderately affected by aging, not nearly as much as the frontal lobes. Hippocampus is found on the inside aspect of the temporal lobe in each hemisphere and is important in the formation of new memories. The amygdala (the word means "almond" in Greek; reflecting its shape) is found right in front of the hippocampus on the inside aspects of the temporal lobes and is important for the experience and expression of emotions.

Interestingly, the hippocampus is not affected by aging in other mammalian species, such as monkeys and rodents. This may be merely a chance difference, but it is also possible that evolutionary pressures favored the human brain with the slightly decaying hippocampus. For those among us with an unbounded

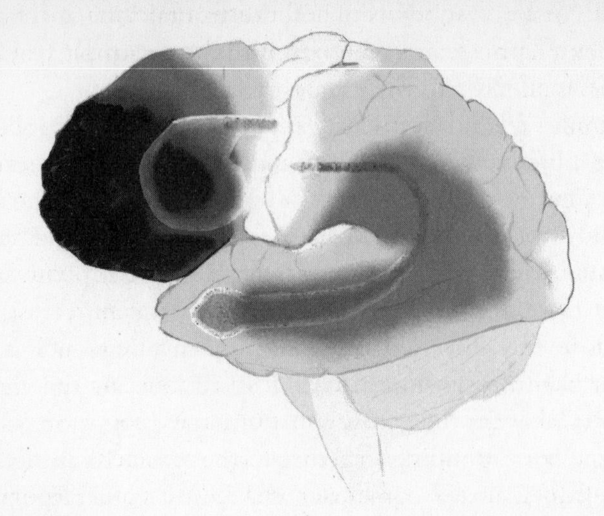

FIGURE 5. **Map of Brain Regions Affected in Aging.** *The darker the color, the more susceptible the brain structure is to the effects of normal aging.*

faith in the adaptive nature of evolution (but prudent enough not to lapse into an outright teleological frame of mind), what could the nature of such evolutionary pressures be? Just as an idle possibility to consider, it could conceivably be related to the fact that humans depend on previously acquired cognitive templates much more than do other species. Hence, an aging human brain, unlike an aging monkey or rodent brain, may benefit from dampening the formation of excess new information, which somehow competes with these templates.

Another interesting finding is the difference in the relative vulnerability of various brain structures in normal aging and in dementia. Unlike in normal aging, in Alzheimer's disease the hippocampus and the posterior heteromodal neocortex of the temporal and parietal lobes deteriorate more rapidly than the frontal lobe. Thus, the disparity between the deterioration of the frontal lobes and the hippocampus evident in the MRI of an aging brain may tell us whether it is undergoing the process of normal aging, or whether it exhibits early features of Alzheimer's disease.

The fate of various subcortical structures generally follows the same Jacksonian principle of "evolution and dissolution." The basal ganglia and the cerebellum (both important for various aspects of motor control) are moderately affected, and so is the midbrain. The pons (the brain area responsible for basic arousal) and the tectum (the first station of sensory input processing within the brain) seem to be affected very little or not at all.

How do these profound changes in brain anatomy translate into the changes in brain function, into cognitive changes? Again, numerous studies have been conducted, painstakingly documenting the adverse mental changes that attend normal aging. It appears that the overall speed of mental operations declines, as do the sensory functions (the ability to receive inputs about the physical world outside). The functions that depend on the frontal lobes appear to falter in particular. These include mental inhibition, the capacity to refrain from distractions or from habitual, knee-jerk reactions to situations. They also include "working memory," a loosely used term employed by most scientists to refer to the ability to hold certain information in mind while engaging in some cognitive processing for which this information is germane. Another function of the frontal lobes, mental flexibility (an ability to switch rapidly from one mental process to another and from one frame of mind to another), has also been found to decline with aging.

Certain forms of attention are also impaired, particularly selective attention (the ability to pick out salient events in the environment and to concentrate on them) and divided attention (the ability to shift attention back and forth among several activities unfolding in parallel). Memory is not spared either. This particularly concerns the ability to learn new facts (semantic memory) and to form memories about specific events (episodic memory). In fact, the erosion of new learning is among the earliest manifestations of cognitive aging.

Ageless Brain

This frightful litany of adverse cognitive changes has been documented by giving the subjects various laboratory neuropsychological tests, and by comparing their performance across age groups. Clearly, the cognitive woes parallel morphological and biochemical woes of the brain, and all this sounds like pretty bad news.

But a closer look at aging cognition leads one to conclude that the news is not as bad as it might appear to be. A puzzling phenomenon did not fail to escape the attention of numerous scientists. Despite this multifarious, well-documented neurological and cognitive decline, it is very common for aged individuals to perform quite competently in real-life situations, both in their everyday lives and on the job. This often includes the discharge of very high-level professional and executive responsibilities, and even world-class feats of artistic and scientific creativity and statesmanship.

Scientists commonly refer to this mysterious ability as "cognitive expertise" and its mechanisms have for years remained obscure. The examination of these mechanisms will be among the focal points of this book. Having faced up to the bad news, it is now time to consider the good news of aging! This mysterious cognitive expertise, which has the uncanny ability to resist the unwelcome effects of aging, resonates with two other highly prized traits commonly associated with mature age: competence and wisdom

There seems to be a paradox there. And since cognitive expertise, competence, and wisdom are not extracranial phenomena hovering over our heads like a saint's halo, but rather very much products of our brain, this paradox becomes a question of neurobiology, a question for a neuroscientist to tackle. In the forthcoming chapters we will examine the phenomena of wisdom and competence and will then proceed with the tour of their neural machinery. But first, let us consider the paradox itself and see how very cogent cognition may be supported by brains touched by aging and neuroerosion. To that end, we will examine the lives of several historical personalities from various arenas of human accomplishment.

3

AGING AND POWERFUL MINDS IN HISTORY

Late-Blooming Achievers

Humans are among the relatively few species with an average life span extending far beyond the age of procreation. Why did evolution contrive (pardon the anthropomorphic and teleological turn of phrase) to prolong the lives of individuals who have nothing more to contribute to the propagation of the species through biological means? What were the evolutionary pressures leading to this odd phenomenon? One possibility is that the elderly make a critical contribution to the survival of the species through other means—particularly through the accumulation of knowledge and its transmission to the new generations via cultural means, such as language. While obvious to scholars, this point has been often overlooked in popular culture.

In our culture, mental vigor is often associated with youth, and mental decline with age. The creative potential of the aged is often dismissed. My friend's nineteen-year-old son Jaan put it in a capsule emblematic of our cultural prejudice: "I am surprised when people your and my father's age are capable of learning anything new at all!" That his father has been among the most compelling educational innovators in Europe, the head of a major university, a presidential candidate, and at the time of this writing a high-profile member of the Parliament in his North European country, seemed to have left the young man utterly unimpressed.

Today, Jaan's dismissive thinking is challenged by numerous examples of successful and yes, innovative people of relatively advanced age—like his father, his father's friend (I like to think), and many, perhaps most, of the readers of this book. This fact is too obvious, too widely accepted and supported by too many examples for me to belabor it at length in this book. By repackaging it slightly and presenting it as an earth-shattering revelation, I would be insulting your intelligence. So I will focus on two less obvious points, which, if anything, amplify the main premise.

My first point is that not only is it possible for a vigorous mental life to continue through the whole lifespan, but also in some people actually reaches its peak at a rather advanced age. I call such individuals *late and luminous bloomers*. History is replete with examples of great creative genius and political leadership reaching its peak only by the age of sixty, seventy, and even eighty. Examples of such remarkable individuals, whose greatest achievements took place late in their lives and became synonymous with their names, can be found in the worlds of literature, architecture, painting, science, and politics. Below are six examples that challenge our well-entrenched cultural bias that aging invariably equals decay.

Johann Wolfgang von Goethe (1749–1832), the great German writer, is clearly a case of an "uphill" life in literature. He published the first part of *Faust* at the age of fifty-nine and the second part at eighty-three. Goethe was a very prolific author throughout his literary career. Nonetheless, it is *Faust,* his late-life achievement, that has been synonymous with his name through the centuries. The life of Antoni Gaudí i Cornet (1852–1926), the great Catalan architectural visionary, followed a similar trajectory. He began the work of his life, the Sagrada Familia Cathedral in Barcelona, exploring architectural forms then without precedent in the Western tradition, as a relatively young man. But the project culminated toward the end of his life, when he focused exclusively on his beloved Sagrada Familia. Gaudí died in a car accident at the peak of his creative powers at the age of seventy-four, and the cathedral remained unfinished. Closer to home, Anna Mary Robertson (1860–1961), better known as Grandma Moses, began to paint only in her seventies. By the time her paintings of rural farming scenes began to gain recognition, she was almost eighty years old. Grandma Moses continued to paint until the very end of her long life and is remembered today as one of the foremost American folk artists.

In a very different arena of human accomplishment, Norbert Wiener (1894–1964) defied his own saying that "mathematics is very largely a young man's game." Wiener was the father of cybernetics. By postulating the existence of unifying principles of complex organization underlying all biological and artificial systems, he shaped much of contemporary science. A unique blend of mathematician and philosopher, Wiener published his *Cybernetics* at the age of fifty-four, and his second most important work, *God and Golem, Inc.*, at the age of seventy. The modern-day science of the general principles governing complex systems, known as "complexity studies," owes much of its foundation to Wiener's insights, many of which were formulated at a relatively advanced age.

Examples of late-life ascendancy to the pinnacle of political leadership are no less remarkable. Golda Meir (1898–1978) served as prime minister of Israel from 1969 through 1974, and guided her country through some of its most momentous crises. She assumed the leadership of Israel at the age of seventy-one, older than Winston Churchill at the beginning of his first term as prime minister (sixty-five), or Ronald Reagan at the beginning of his first presidency (sixty-nine). She was known toward the end of her life as the "Mother of Israel." Nelson Mandela (1918–), one of the most compelling political personalities of the twentieth century, served as the first democratically elected state president of South Africa from 1994 through 1999. Mandela assumed the presidency at the age of seventy-six, his clarity of mind and force of personality undiminished by a twenty-eight-year imprisonment. Mandela helped mold his country's new identity in definitive ways, and he remains the symbol of free South Africa at the time of this writing.

One might say that the late-life creative accomplishments, and even a late-life creative peak, exemplified by our six examples, are merely a matter of genetic luck, that some people are fortunate to retain their mental acuity well into old age. While encouraging, such examples are not particularly surprising, since every curve has its outliers. But now we are ready to reach yet another, truly unexpected conclusion, which brings us to my second point.

My second point is that even partial loss of mental powers does not necessarily portend "cognitive doom"—that a person may remain productive and cognitively competent in important ways, even despite measurable cognitive decline, perhaps even despite early dementia. I call such individuals *eroding yet powerful minds.* The thought of a person at an early stage of a dementing process being able to make important contributions to the cultural or political life of society may sound at the first blush outlandish, but careful examination of history leads to this astounding discovery. Some of the most fateful political decisions (both constructive and destructive) and lasting artistic creations were made by minds touched by well-documented neurological effects of aging, sometimes even by early dementia. This is true both in politics and in the arts, and possibly in philosophy and science as well.

The account of our history and culture being influenced by individuals at various stages of neurological decline and early dementia makes for amusing reading. But merely recognizing their mental infirmities distracts us from a much more interesting question: What were the attributes of their minds that compensated for the effects of neurological erosion and preserved their mental power and effectiveness, their ability to shape culture or politics and to dominate their worlds? To a great extent, the compensation was provided by a rich arsenal of pattern-recognition devices, which had been formed in their brains decades earlier.

The etymology of the word "dementia" is "the loss of mind." It is a cruel, merciless, doom-spelling word. It implies a certain, rather significant amount of cognitive loss. It has threshold connotations. For all these reasons the term "dementia" should be used sparingly. In reality, most forms of dementias develop gradually and in fact rather slowly. The decline extends over years, sometimes as long as a decade and a half, and in some isolated instances even longer. It is not as if a precipitous transition from total lucidity to total mental blackout were to take place overnight; far from it. Nor is it true that dementias affect all the mental faculties at once. In most cases, the process first affects only certain faculties, while others remain spared for a while, often for long periods of time measured in years. But ultimately the disease spreads. During the early stages of the process, the afflicted individual is still in command of most of his or her mental facilities and may be fit to perform complex activities, even highly intellectual ones, for a number of years to come. While such a person may be at an early stage of a downhill slope leading eventually, and in many cases inexorably, to full-blown dementia, he or she is not yet nearly demented and will not be for years. Furthermore, not every case of mild cognitive impairment will progress toward a full-blown dementia. So, there is a difference between a dementing process and down-and-out dementia. This fact has been long recognized by physicians and psychologists, and different stages of mental decline have been elaborately described.

Earlier I made the point that a mind equipped with a wide range of previously formed pattern-recognition devices can withstand the effects of neuroerosion for a long time. In the forthcoming chapters we will discuss the brain mechanisms ensuring such protection. But first let us examine the phenomenon itself, just so there is no doubt in the reader's mind that it is possible to be neurologically affected by aging and cognitively powerful at the same time, implausible as this may sound.

On the pages to follow, I will discuss the lives of several remarkable artists and political leaders who were cognitively affected by aging while making an indelible imprint (for better or for worse) on history and culture. I will talk about their neurological infirmities and early signs of mental decline intertwined with impressive accomplishments. We will begin with the lives of two of the twentieth century's greatest artists.

Art and Dementia

The Basque country, straddling the Spanish–French border, has long been regarded a land of mystery. The Basque language is itself unique, unlike any Indo-European language, its origins uncertain. The Basque people are presumed to represent the earliest population of Europe, related to the Celts or possibly even pre-Celtic peoples, a vestige of the tribes inhabiting the continent before the multiple waves of migration and conquest changed its ethnic and linguistic complexion. The Basque country is also more recently known for its volatile, sporadically violent independence movement, although for a tourist this is an abstract notion, and there is no palpable feeling of menace in the air. Quite the contrary, the Basque provincial capital San Sebastian is among the most famous European beach resorts, synonymous with boats, sun, excellent restaurants, and the sybaritic pursuit of the good life. The area is also the home of a unique tradition of monumental sculpture associated in particular with the names of the great Basque sculptor Eduardo Chillida (1924–2002) and his lifelong rival Jorge Oteiza (1908–2003).

During my visit to San Sebastian, the conversation over dinner turned toward Chillida, who had died earlier that year at the age of seventy-eight. My hosts, neurologists from the local medical center, were recounting how the famous sculptor had ended his life in their care, in a state of advanced Alzheimer's dementia. It turns out that Chillida was completely incapacitated during the last year of his life, his mental powers sapped by the disease.

The next morning we drove to the famous Museo Chillida-Leku, a sculpture garden in the nearby village of Zabalaga, which houses the largest collection of works by Chillida. The vast estate is centered on a sixteenth-century barn, converted by Chillida into a residence and surrounded by lush gardens and lawns studded with sculptures. Chillida's work is monumental and mostly abstract. He used metal, marble, stone, and wood to create nonrepresentational yet highly evocative shapes, a magical fusion of a Cyclopean scale and introverted private moods. As I was strolling among the gigantic forms, I felt that an elusive similarity existed between these contemporary sculptures and Stonehenge. They seemed ageless, inspired by the same muse, or at least by the same lineage of muses. The Basques and the Celts are both direct heirs of the ancient peoples of Europe, pushed to the westernmost fringes of the continent by the invading waves of newcomers. Could it be that their shared history translated into shared artistic sensibilities, transcending the four millennia separating the druids of Stonehenge from the Basques of today, that an ancient tradition found its modern-day expression in the works by Chillida and Oteiza? The thought amused me and created a pleasant buzz in my head as I continued my stroll through the sculpture garden.

And then I began to notice that some of the plaques next to the sculptures, in fact quite a few, bore the dates in the mid-nineties, late nineties, and even the year 2000. As we already know, Alzheimer's disease does not assault one all of a sudden. Quite the contrary, it is a gradual decline, a slippage into mental oblivion that unfolds over years, not months. Someone who was in a state of advanced dementia in 2001, as reportedly Chillida was, certainly had to be already affected by the disease process in the late nineties, and probably as early as in the mid-nineties. Yet here I was surrounded with the masterpieces, which every curator of every major museum in the world would give an arm and a leg to have . . . created by an artist most likely suffering from Alzheimer's disease. When I shared my chronological observations with my hosts, they seemed as perplexed as I was. We left it at that, but the image of an aging master, losing his memory but not the secrets of his craft and triumphing over his illness through his art, at least for a while, kept haunting me for months after the visit.

Eduardo Chillida and his poignant story find a counterpart in a North American contemporary and fellow artist, Willem de Kooning (1904–1997). A Dutchman who came to the United States in 1926 at the age of twenty-two and made it his home, de Kooning epitomized twentieth-century American art like no one else. His career as a painter and occasional sculptor spanned three quarters of a century. De Kooning was a true original who helped forge a new direction in painting. Being an original was the essence of his identity. "Nothing grows under big trees," he once told a student who was quizzing him as to why he had never studied with a famous artist. He himself became that "big tree," which in defiance of his own admonition spurred the growth of a whole new school. From an early infatuation with cubism, through the transitional stages of painting, by his own account, increasingly abstract "quiet men" and then "wild women," de Kooning moved on to become a founder of what has since become known as "abstract expressionism."

Sometime in the late 1970s, de Kooning's memory loss became evident to those around him. As is usually the case, his amnesia affected his memory for relatively recent events and spared the memories of the distant past, a phenomenon well-known to neuropsychologists and neurologists under the cumbersome name "the temporal gradient of retrograde amnesia." But even more distant memories may have faded as the disease progressed. His biographer Hayden Herrera recounts an episode in which de Kooning was unable to recognize an old and close friend of many years. The diagnosis of Alzheimer's disease eventually followed.

But the old master continued to paint, spending all his days in the studio, sometimes finishing several paintings a week. "A finished painting is a reminder of what not to do tomorrow," he was quoted to quip at the age of eighty-one. (His memory may have eroded, but his wit was undiminished.)

De Kooning's art continued to evolve even toward the end of his career. In the 1980s his brushstrokes broadened and then toward the late 1980s his paintings began to acquire what his biographer and friend Edvard Lieber called "hyperactive forms"—spare, brightly colored, wavy curves. De Kooning, well into his eighties, was aware of the change: "I'm back to a full palette with off-toned colors. Before it was about knowing what I didn't know. Now, it's about not knowing what I know." This change was more than a change in style. For de Kooning, his work had always been a means of comprehending a deeper meaning of things and of his own experience, and not merely forging a set of formalisms. "Style is a fraud. . . . To desire to make a style is an apology for one's anxiety," de Kooning wrote many years earlier.

So what evolution of de Kooning's own human experience did the changes in his work reflect? What role did the change in his cognition play in the evolution of his art? Was the effect one of decline or one of ascendancy? Or some complex interplay of both?

The change in de Kooning's work did not elude the art critics. It was regarded as evolution and not as regression, as the ascendancy to a new level of insight and understanding. "The rhythms are more deliberate, meditated even, and the space more open . . . a new order prevails, a new calm. . . . de Kooning has purified his stroke, and what had been quintessentially sensuous is rendered immaterial, ethereal, a veiled tracing of its physical origins," wrote David Rosand. "de Kooning, who has never strayed far from nature for long, is closer to it now than ever," wrote Vivien Raynor in the *New York Times*.

So here are the stories of two great twentieth-century masters, Eduardo Chillida and Willem de Kooning, who were able to create first-rate art despite the progression of Alzheimer's disease, with its crippling effects on many other aspects of their lives. Before we proceed further with the discussion of what made this possible, let us step back and appreciate the sheer power of the facts themselves, whatever their explanations may be.

Leadership and Dementia

To fully appreciate the force of these facts, let us also make note of their universal nature. The art is not the only arena in which the masters of their crafts retain their touch despite the crippling effects of the assorted brain diseases of aging. Let us also consider the arena of statesmanship and politics. And here we are stepping into a morally agnostic territory. If the great artists are remembered for their good, at least as public personas, then the important statesmen and politicians can be either heroes or villains, or tangled juxtapositions of both. We will consider examples of all of the above among those aspiring to rule despite their cognitive decline and even early dementia.

"First among the virtues found in the state, wisdom comes into view," wrote Plato in his *Republic*. We wish! We often think of the rich and the powerful as exempt from the laws of nature, including the laws of physics and biology. What's more, the rich and the powerful are probably the first to share this belief. This is benevolently known by some as "boundless self-confidence" and less benevolently by others as "hubris."

But whatever may or may not be true for other natural laws, the biological processes causing dementia do not discriminate on the basis of wealth, power, or even moral rectitude. We are only beginning to understand dementia's biological causes and the processes by which it robs the mind of its powers and turns the most brilliant intellect into a shell, an incoherent and confused wreckage of a human being. Many forms of dementia exist, some causing gradual brain atrophy and others causing a gradual accumulation of small strokes. To make matters worse, they often appear in combinations. All dementias are equal-opportunity scourges, eroding the mind in a variety of insidious ways, without sparing the rich, the powerful, and the righteous. It is amazing how many history-shaping decisions have been made, and continue to be made, by eroding, even dementing minds before the eyes of a power-awed, unsuspecting public.

This thought first crossed my mind many years ago, as I was making my diagnosis of Ronald Reagan. A refugee from the former Soviet Union, I had been an anomaly among my friends in the liberal New York intelligentsia as an admirer of Reagan, the man who helped dismantle the "evil empire" I had fled half a lifetime ago. So, when the inkling of Reagan's dementia first crossed my mind, I was far from gloating; I was genuinely upset. That was well before Reagan's Alzheimer's disease became public knowledge or even a matter of public speculation. In fact, it was well before Reagan left the White House.

Sometime during his second term, Reagan was quizzed by a journalist about the wreath-laying Bitburg affair, when in 1985 Reagan honored a cemetery full of Nazi SS guards against the advice of his aides. The feeling was that the American president was being manipulated by the then West German chancellor Helmut Kohl, who needed the gesture for his own political ends. As I was watching the interview on TV, Reagan's responses to the journalist's questions sounded so staggeringly incoherent that I picked up the phone, called my neurosurgeon friend (and a fellow foreign affairs buff) Jim Hughes, and said: "Reagan has Alzheimer's!" Jim laughed, not realizing that I meant it literally, and not as a figure of speech.

This may have sounded like a snap judgment, gratuitous even, but I was better equipped for making it than most people. A neuropsychologist with (then) almost twenty years of clinical experience and a reputation for diagnostic acumen, I make a living by studying, diagnosing, and treating various brain diseases affecting the mind. I was also doing research, publishing scientific papers, and writing books about the brain and the mind, and the numerous ways in which they may go wrong. The incoherence that so struck me in Reagan's responses would have raised my diagnostic antennae coming from anyone, and Ronald Reagan was not exempt.

My hunch about Reagan was strengthened some time later, during the last day of his presidency, as I was watching George Bush's inauguration on TV. Reagan walked past the honor guard, approached the imposing leather chair prepared for him, slumped into the chair, and was immediately asleep, his head dropping on his chest instantaneously. "Brain stem gone," I said to myself, alluding to the part of the brain that is in charge of maintaining the arousal necessary for sound mental activities. At this point I was convinced that a significant portion of Reagan's second term had taken place in the shadow of his slippage toward early dementia.

My conclusion that Ronald Reagan was suffering from Alzheimer's disease or a similar dementing condition was sealed soon after he left office and well before the first official intimation to that effect. As I was watching Reagan's interviews about the Iran-Contra affair, I was impressed, almost shocked, by the sincerity of his denial of any memories of the events, by the befuddled and incredulous expression on his face when the events and names of people were being thrown at him by the interviewers. Contrary to the opinion of many commentators, I was convinced that Reagan was not dissembling, that he was not attempting to hide anything. With the confidence of an experienced clinician, I felt that he truly did not remember. Ronald Reagan was suffering from early dementia.

Of course, my diagnosis via television was subsequently confirmed when the "official" diagnosis was made in 1994 at Mayo Clinic, and Reagan's hereditary risk factors revealed (both his mother and older brother had suffered from dementia). The former president's own courageous admission of his illness earned him my respect and that of many other people. Were my earlier observations of Ronald Reagan indicative of outright dementia, or did they still belong in the gray area of "neuroerosion" or "mild cognitive impairment," the early prodrome of things to come? Ultimately, this is a matter of semantics more than of substance, since we are talking about a gradual downslide devoid of discrete boundaries and not about an abrupt transition, a decline that came to an end in 2004, ten years after the "official" diagnosis of dementia and considerably longer after it had actually begun to set in.

Heroes and Villains

My clinical television study of Ronald Reagan brings us to a much broader issue. His case is certainly not unique. The paradox of human society is that the age of ascendancy to the summit of power in our political, cultural, and business institutions is also the age of onset of numerous forms of neurological decline. A large number of world political leaders are men and women in their sixties and seventies. And while we accept as a given the inevitability that by this age assorted physical infirmities accumulate, society is by and large oblivious to the fact that by this age dementia also develops in a significant number of people.

The illusion that the demigod figures who make it to the summit of human society are spared the indignity of brain rot is precisely that: an illusion. Dementia operates on the basis of age and genetic vulnerability, just like any physical malady. Dementia is an age-related physical malady affecting the brain, just as coronary insufficiency is an age-related physical malady affecting the heart. The mind is not exempt from the fundamental laws of biological decay.

One might expect that the individuals who make it to the very top are brighter than the population as a whole, and this is probably mostly true. But history is replete with instances of individuals endowed with great intellectual powers succumbing to dementia toward the end of their lives for reasons of genetics, or for some yet-to-be understood environmental reasons. Contrary to our wishful thinking, an exalted social status does not offer protection in these matters, nor, as it turns out, does a great intellectual power.

It is intuitively plausible, and certainly teleologically "desirable," that great minds should be protected from decay. Indeed, the last decade has witnessed a paradigm shift in neuroscience, as the evidence began to accumulate that vigorous mental life reshapes the brain itself and helps protect it from biological rot. (Much more about this later in the book.) But other factors, like heredity, are less malleable, at least today.

The history of science and philosophy is similarly replete with poignant observations of decaying great minds. Isaac Newton, Immanuel Kant, and Michael Faraday all suffered dramatic memory loss with age. Among the more recent examples, Claude Shannon, the father of information theory, was diagnosed with Alzheimer's disease toward the end of his life.

But mental decline in a scientist is not likely to result in a societal disaster. It may have a retarding effect, delaying a great discovery or invention by years, decades, or even generations, but hardly a precipitously catastrophic one. Besides, most great scientists have their definitive insights relatively early in their careers. By the time dementia is likely to strike, they will have already made their seminal contribution to society long ago, and their decline, sad as it may be on a personal level, is no longer of broad historical relevance.

Not so with a political leader, a powerful statesman at the

helm of a major military or state machine, when the age of supreme power often overlaps with the age of early cognitive decline, under whose shadow fateful decisions are made. Mental infirmity may take many forms, from what I call mild "neuroerosion" to frank dementia, but the brain machinery of the sublime and the ridiculous is fundamentally the same. A world leader whose decisions affect the lives (and deaths) of thousands of people fundamentally employs the same brain machinery as the owner of a family-run neighborhood bodega making a decision about what brand of canned tuna to stock next week. This means that the consequences of an early "mild" dementia, which may be imperceptibly benign in a neighborhood grocer, will be perilously magnified in a world leader through the sheer impact of his mental faux pas.

Reagan was in his seventies at the time of my observations. At this age, Alzheimer's type dementia, multiinfarct dementia (a disease of blood vessels of the brain resulting in a multitude of small strokes), and other forms of dementias are all distinct statistical possibilities. An early-stage dementing disease process often eludes detection by an untrained eye even in a leader who is constantly in the public eye. But it is particularly likely to remain unnoticed or ignored under the conditions of an authoritarian regime, where the leader is relatively exempt from popular scrutiny. Impairment of judgment, self-control, and other higher mental functions, first subtle and then increasingly transparent, takes place well before an individual becomes frankly disoriented, totally disabled, and no longer capable of hiding his mental infirmities even from distant observers.

The past century witnessed the stewardship of more than a few "neuroeroding," dementing, or indeed demented individuals at the helm of major nations. Dementia strikes the villains and heroes of our world in a morally agnostic way.

On the villains' side, Adolf Hitler suffered from severe symptoms of Parkinson's disease toward the end of World War II. According to some reports, memory decline was also apparent.

Contrary to the popular belief, Parkinson's disease is not just a movement disorder. It often causes some degree of cognitive impairment and even outright dementia. Other conditions also exist whose outward symptoms resemble those of Parkinson's disease, but in which serious mental impairment is expected to be present. Most common among them is Lewy body dementia, a degenerative brain disease of aging. At the age of fifty-six, toward the end of the war, Hitler was more likely to suffer from Parkinson's disease than from Lewy body dementia. Either way, based on simple epidemiological considerations, some mental deterioration was highly probable. Indeed, his close associate, Albert Speer writes about Hitler's "apathy," "mental torpor," and difficulties with decision-making becoming increasingly evident during the second half of the war.

The other great villains of the twentieth century were not spared either. During the last years of his life, Joseph Stalin, known for his extraordinary memory in earlier years, was reported to have memory lapses, even forgetting the names of close associates. There was a notable exacerbation of Stalin's paranoia (a common symptom of dementia), and it became even more dangerous for those around him than before. His lieutenants "were convinced that Stalin was becoming senile," according to Simon Montefiore. After the war, Stalin "wasn't quite right in his head," Nikita Khrushchev is quoted as saying, an impression shared by the visiting Yugoslav communist Milovan Djilas. Stalin's command of Russian (not his native language, but one in which he had attained remarkable facility) deteriorated, and he had difficulties expressing himself. The loss of command of a second language, and reversal to the language of one's childhood (in Stalin's case Georgian), is a well-documented consequence of dementia in bilingual individuals. Stalin also suffered from transient episodes of disorientation and dizziness common in cerebrovascular disease. Montefiore further writes that in the spring of 1952 Stalin was examined by "his veteran doctor" Vladimir Vinogradov, who concluded that

Stalin suffered "minor strokes and little cysts in the brain tissue of the frontal lobe." The autopsy of Stalin's brain, conducted in 1953 after his death of a stroke (or, as some historians believe, of poison) showed signs of arteriosclerosis of at least a five-year duration. Today his condition would be called "early multiinfarct dementia."

Stalin's mentor Vladimir Lenin, arguably a villain in his own right, also suffered from multiinfarct disease of the brain (a complication from chronic syphilis infection, according to some historians). He had a series of debilitating strokes between 1922 and his death in 1924, and lost much of his use of language. Yet he continued to run the nascent Soviet state intermittently, between successive strokes, until 1923, while already undoubtedly cognitively impaired.

Mao Zedong's eccentricities toward the end of his life have been described as well. He was known to suffer from amyotrophic lateral sclerosis (ALS), a neurodegenerative disease, characterized by the death of motor neurons. This disorder, also known as Lou Gehrig's disease, causes progressive loss of movements, including the control of one's motor apparatus of speech. Toward the end of his life, Mao's ability to communicate by means of language was so impaired that his speech became virtually unintelligible. But this may not have been the whole story. Contrary to previous neurological beliefs, the symptoms of ALS are not limited to motor difficulties. Recent research has shown that significant cognitive impairment, including outright dementia (affecting particularly the frontal and temporal lobes, where higher-order processes such as decision-making and language are based), is present in more than a third of ALS patients. This cognitive impairment affects mental flexibility, abstract reasoning, and memory.

Yet despite their mental infirmities, Hitler, Stalin, and Mao all remained at the helm of their respective "evil empires," as Alan Bullock points out, until the very end of their lives, compounding their lifelong propensities toward villainy with mental deterioration or outright early dementia.

The brain diseases of aging did not spare the political heroes of the twentieth century either. Woodrow Wilson suffered a severe stroke while in office in 1919. He recovered, but only partly. According to his biographers, Wilson was a different man after his stroke. His mind became rigid, devoid of nuance, casting everything in black and white. These newly acquired untoward traits haunted the last two years of his presidency and undermined his ability to deal with the isolationist Congress, contributing to the ruin of his League of Nations policy.

Franklin Delano Roosevelt was felled by a lethal stroke, but a major stroke is often preceded by what is known today as multiinfarct disease, characterized by a gradual accumulation of ministrokes. In FDR's days this condition was not known, nor were there any diagnostic tests available capable of revealing it (such as a CT scan or an MRI). Nonetheless, the decline of FDR's mental powers and decision-making abilities, and his "new disinclination to apply himself to serious business" during the final phase of World War II, have been noted by credible historians. He was likely suffering from cognitive decline already well before his final stroke.

And so was the man whom I admired more than virtually any other twentieth-century political leader, Winston Churchill. When he was elected to his first term as Britain's prime minister, Churchill was already sixty-five, older than most of the last century's other major political leaders at the time of their ascendancy to supreme power.

Churchill's occasional mental lapses during World War II have been noted by both his wartime associates, like Field Marshal Alanbrooke (leaving them occasionally worried about their leader's mental state), and his biographers, like Roy Jenkins. These lapses nonetheless did not prevent him from dispatching his business with overall brilliance, if with only occasional flagging. Churchill suffered his first known minor stroke in 1949, between his two terms as prime minister. During his second, post-war term in 1951–1955, Churchill was, in the memorable words of Roy Jenkins (as sympathetic a biographer as any public figure can hope for), "gloriously unfit for office."

According to the accounts of those around him reviewed by Jenkins, Churchill's energy level during his second prime-ministerial term went precipitously up and down, and so did his powers of concentration, speechwriting, and ascertaining complex ideas. He spent an inordinate amount of time playing the esoteric card game of bezique. He suffered a succession of several minor strokes. Then, in 1953, while still in office, Churchill was felled by a severe stroke and for a period of time remained wheelchair-bound, his speech slurred. By coarse neurological standard, he recovered well, but was no longer his old self, and those around him awaited, with a mixture of reverence and impatience, his resignation, which was not readily forthcoming, as he used every excuse to postpone it until April of 1955.

More recent political history is also replete with examples of mental decay in political leaders while in office. Leonid Brezhnev, the leader presiding over the "stagnation period" of the former Soviet Union, was on many occasions toward the end of his rule less than totally coherent, his speech slurred and his gait unstable. Dmitri Volkogonov, a noted Russian historian and a three-star general close to the upper strata of the Soviet leadership, desribes Brezhnev's demeanor during his last years in office as "senile and confused." Reagan's friend and Churchill's illustrious Tory Party successor Margaret Thatcher announced her departure from public life due to a succession of "mild strokes," and it sounded as if Lady Thatcher had been suffering from an early stage of this cognitively debilitating disease. Unlike the constraints imposed on the American or French presidency, there is no constitutional limit to the number of terms served by a British prime minister. Under a different set of circumstances, the Iron Lady may have continued to prevail again and again, and her tenure as the leader of Europe's oldest democracy would have overlapped with the onset of an insidious dementing condition.

The last decade of the twentieth century saw more such

examples. Former presidents Boris Yeltsin of Russia and Abdurrahman Wahid ("Gus Dur") of Indonesia are the two more recent cases of dementing leaders at the helm of some of the world's largest nations. Yeltsin was a clinical alcoholic and a heart patient, probably suffering the irreversible changes in the brain common to these conditions. Any head of a major state who urinates on a foreign airport tarmac in front of a receiving line of dignitaries has to be more than merely drunk. Abdurrahman Wahid of Indonesia, one of the transitional figures following the deposing of Mohamed Suharto, suffered several major debilitating brain-damaging strokes. He was notorious for his less-than-coherent ramblings.

The stewardship of each of these two leaders to their respective countries was a mixture of good and bad. Both were known for their erratic, contradictory, and often incoherent behavior, an odd reflection of the transitional nature of their leadership. I strongly doubt that either Yeltsin or Wahid, or for that matter, Brezhnev would have passed a standard neuropsychological dementia evaluation commonly administered in North American geriatric clinics.

This review of mental infirmities in world leaders adds up to a rather staggering picture, particularly in light of the recent revisions of what constitutes "normal aging" and what does not. Among past generations, cognitive loss, "losing one's marbles," or "being out of it" was considered an integral and normal part of aging. We no longer think so. In their groundbreaking book *Successful Aging*, John W. Rowe and Robert L. Kahn challenged the notion that mental decline is either normal or inevitable, and argued with great force that mental decline in aging is due to one or more identifiable diseases of the brain, many of them potentially preventable or treatable. They introduced the notion of "successful aging," which among other things includes complete lucidity and mental sharpness well into old age. Rowe and Kahn argue that this, and not mental decline, is the norm. Those sprightly, astute, quick, and mentally nimble octo- and nonagenarians, such as Federal Reserve Chairman Alan Greenspan or celebrated historian Jacques Barzun, are now my role models. I always wonder whether I will be even remotely like them in my own old age, should it come to that at all.

The rub is that some of the most important twentieth-century world leaders did not seem to have enjoyed successful aging, as far as their brains were concerned. Quite the contrary, from a neurological standpoint the towering personalities who dominated the twentieth century's political landscape, both heroes and villains, aged appallingly badly.

While the historical anecdotes assembled in this chapter make for entertaining reading, it is important not to miss the main point: Despite their often significant mental infirmities, most of these leaders remained in control. Although undoubtedly shielded by layers and layers of aides and secretaries, most of them, both heroes and villains, continued to be at the helms of their countries as real leaders, and not as mere figureheads. Most of them were on top their respective games almost until the very end. While staggeringly implausible at a first glance, this has been true on numerous occasions throughout history. As we have already seen, a number of great cultural personalities were able to maintain their artistic acumen despite significant cognitive erosion, even dementia.

What allowed these remarkable personalities to prevail despite neurological decline was the rich, previously developed pattern-recognition facility, which enabled them to tackle a wide range of new situations, problems, and challenges, as if they were familiar ones—an advantage that those around them and those opposed to them lacked. The remarkable personalities described in this chapter are cases in point to Herbert Simon's claim that pattern recognition is the most powerful cognitive tool at our disposal. Their histories show with dramatic clarity that the machinery of pattern recognition can withstand the effects of aging on the brain to a remarkable degree; that the protection offered by this machinery to an aging mind can be nothing short of profound; and that the empowering effect of a well-developed arsenal of essential patterns stored in one's mind can remain intact at very advanced stages of life. The machinery of pattern recognition can withstand even the effects of age-related dementias to a considerable degree and for a long time.

Not all of the important personalities described in this chapter attained wisdom—far from it—but it can be argued that all of them exhibited expertise and competence within their respective cognitive arenas, some good, some evil. They may have lost some, and often much, of their mental computational power. Their memory and attention may have been significantly affected. But they had accumulated through their previous experience a large number of cognitive templates. This enabled them to tackle a wide range of complex situations as familiar patterns, despite their mental erosion, and to dominate, for better or worse, their more computationally nimble but less "pattern-recognition enabled" colleagues, associates, and most importantly, adversaries. Exactly how the cognitively enabling patterns are formed and what protects them from the erosive effects of mental decline will be the subject of the chapters to follow. But first we will examine wisdom, competence, and expertise as psychological phenomena.

4

WISDOM THROUGHOUT CIVILIZATIONS

Wisdom and Genius

Is wisdom a gift or a well-earned prize?

The phenomenon of wisdom has awed generations of philosophers, psychologists, and the general public alike. Its special status was first recognized early in history, and the admiration for wisdom permeates every culture and every slice of civilization, which is captured in the teachings of Confucius and the aphorisms of Solomon. In recent times, leading scientists and public figures have tackled the subject of wisdom as a psychological and social phenomenon. This has resulted in several books sharing the title *Wisdom* but approaching the mysterious phenomenon from vastly different vantage points.

Among them is a particularly informative and lucid collection of rigorous scientific essays summarizing the research conducted by a number of leading scientists and edited by a highly respected Yale psychologist, Robert Sternberg. I found this volume especially helpful in researching for my own book, and many of the facts and insights contained in the essays are reviewed here.

A very different perspective is offered in a book bearing the same title by acclaimed Australian radio journalist Peter

Thompson, who attempted a glimpse into the mysterious phenomenon of wisdom by interviewing several prominent public figures from various walks of life, presumably endowed with the precious gift.

It has always been accepted that of all the mental powers wisdom is the most coveted: *Wisdom is the principal thing; therefore get wisdom* (Proverbs 4:7). But how? And what exactly is it? On a personal level, the sense of attaining wisdom is a source of deep satisfaction and fulfillment. "Wisdom is the supreme part of happiness," wrote Sophocles in *Antigone.* Psychologists Mihaly Csikszentmihalyi and Kevin Rathunde concluded that among "concepts relating to the evaluation of human behavior" wisdom has attracted the most enduring interest throughout the millennia of recorded history. They further say that, although highly intuitive, the concept of "wisdom" has been infused with a certain continuity of meaning throughout more than twenty-five centuries. Psychologists James Birren and Laurel Fisher link the first mention of wisdom to even more remote beginnings of history. They quote the *Encyclopaedia Britannica* as tracing it to ancient Egyptian writings composed almost 3000 BC, noting also the first mention of a man reputed for his wisdom 600 years later, a vizier at the pharaoh's court by the name of Ptahhotep. In more recent times, the Wisdom Tree, with its seven branches of knowledge capped by wisdom, became one of the most emblematic images of medieval Western art, and the Eastern tradition of *The Seven Pillars of Wisdom* was made famous in the English-speaking world by T. E. Lawrence. To this day, we regard order and enlightenment as the celebration of wisdom, and chaos and excesses as the result of a lack of wisdom. Throughout the history, wisdom has been understood as the fusion of the intellectual and moral, spiritual and practical dimensions.

But despite this abiding interest in the phenomenon of wisdom, despite the fact that the nature of wisdom has been debated since antiquity, it remains shrouded in mystery even today.

Until recently no serious attempt was made to understand the brain mechanisms of wisdom and little to nothing had been said or written on the subject. "To understand wisdom fully and correctly probably requires more wisdom than any of us have," says Robert Sternberg. A noted psychologist and a distinguished student of the subject, he should know.

How to approach this seemingly impenetrable subject? An old professor of mine, the distinguished psychologist and a great aficionado of elegant parable, Alexei Leontyev, used to say that in order to make things easier to understand, you first need to complicate them. We will follow this provocative recipe. To this end, as if wisdom were not intractable enough, we will also consider genius.

Wisdom and *genius* are often invoked in the same breath. In fact, Sternberg puts "wisdom" and "creativity" together in the title of his seminal paper. But the nature of genius (or creativity) is as inexplicably mystifying as the nature of wisdom, if not more so. "From remote antiquity until the dawn of what is taken to be modern philosophy, wisdom, like genius, was explicated in terms of providential gods, muses, astrological forces, a sixth sense, genetic bounty, or accidents of nature," Robinson writes. Genius is among the most revered, yet unattainable, human traits, and so is wisdom. Both are the assets of the few and most of us do not pretend, or even aspire, to have either.

Genius and wisdom share the paradox inherent in their being extreme manifestations of the human mind. They are likely to exist among us unnoticed. The paradox is that both genius and wisdom may lead to conclusions so out of sync with the concepts and beliefs prevailing in society at the time that they are discarded as madness or even completely ignored, like babble in a foreign tongue.

The corollary of this paradox is that in order to make an impact, both genius and wisdom must be ahead of society, but not so far ahead as to be incomprehensible. They must challenge the prevailing beliefs and connect with them at the same time.

Military historian J. F. C. Fuller wrote, "Genius can be baffling." By definition it is. But not too baffling, lest it be ignored or laughed at as foolishness. This fine balance was captured by William Wordsworth, who wrote: "Never forget that every great and original writer, in proportion as he is great and original, must himself create the taste by which he is to be relished."

Being too far ahead of one's time is probably more the fate of genius than of wisdom. After all, we can define wisdom as the ability to connect the new with the old, to apply prior experience to the solution of a new problem. But we define genius as the ability to reveal and grasp undiluted novelty in its purest form. Genius too far ahead of its time is likely to be ignored by its contemporaries, and so is likely to be lost to the generations that follow, although it is difficult to blame society for this neglect. "The very essence of the creative is its novelty, and hence we have no standard by which to judge it," said psychologist Carl R. Rogers.

Does this mean that the celebrated minds, the cultural icons, the great scientists and philosophers, whose theories and discoveries propelled the forward thrust of civilization and illuminated its course like beacons in the night—Aristotle, Galileo, Newton, Einstein—were in fact second-best intellects, what a wine aficionado would call "second growth"? That our history has been punctuated with forgotten "extreme geniuses," whose names and ideas have been forever lost to a society incapable of grasping them in their own time? This thought has intrigued, amused, and disturbed me for some time, not the least because of its vaguely blasphemous cultural implications of rejecting the truly best and embracing the second best. But to think about it further, the whole proposition is a bit oxymoronic, since if their names had been forgotten centuries ago, how can we know today that these geniuses ever existed?

Yet sometimes a nearly forgotten genius's memory is salvaged for history by serendipity, a coincidence, a fluke, or through a cultural historian's hard work. I call this the "Leonardo phenomenon." Today, Leonardo da Vinci is recognized as a genius of the first order twice over: a genius artist and a genius inventor and engineer. It was his artistic genius that secured his immortality and thus sustained an enduring interest in every other aspect of his legacy, including the engineering designs in his Codices. But let me ask this question: Had there been no Leonardo the genius artist, and only Leonardo the genius engineer had lived, would we know his name today? I think not. His engineering ideas were so far out, so ahead of their time, that the likelihood of their making an impact on his contemporaries was extremely remote. The memory of Leonardo the genius engineer would probably have been irretrievably lost, had it not been salvaged by Leonardo the genius artist! But the image of a sage scorned and laughed at by shortsighted contemporaries is also not unheard-of. The life of "a prophet without honor in his own land" is known to be the fate of sages too, not just of daring geniuses. Call it the "Cassandra phenomenon" if you will. Think of Mohandas Gandhi beaten by police in South Africa, or Andrey Sakharov sent into internal exile in the Soviet Union.

What is the meaning, if only metaphorical, of the phrase "touched by God?" (As an agnostic with atheistic leanings, I nonetheless use it myself whenever I encounter an individual of unusual gifts.) Do these rarefied traits, genius and wisdom, stand completely apart from the makeup of us simple mortals? If so, what are we doing here in trying to comprehend the incomprehensible, trying to define genius and wisdom, although we may lack even the ability to recognize the true geniuses and sages among us? And how can we relate these demigod-like gifts, wisdom and genius, to the lives and realities of intelligent, but let's face it, ordinary human beings, like most readers of this book, as well as its author?

Are those endowed with wisdom or genius fundamentally and inherently different from us? Are they made from qualitatively different stuff, so to speak, as Michelangelo's marble statue of David on a pedestal is qualitatively different from the throng of admiring flesh-and-blood tourists gaping at it? Or is there continuity between these coveted but mostly unattainable traits and some more modest attributes that many of us have, or at least can realistically aspire to? In other words, could it be that both wisdom and genius are extreme, supreme forms of some highly desirable, but far more common traits? By uncovering such continuities, we will move a step closer to unraveling the mysteries of genius and wisdom. And by identifying and examining the underlying mental traits we will make these concepts more relevant to the lives of most people, who may be both gifted and intelligent, but are neither geniuses nor sages.

Talent and Expertise

To this end, let's consider two highly desirable but less Olympian qualities: talent and expertise. Suppose that *genius* is an extreme form of *talent,* and that *wisdom* is an extreme form of *expertise* or *competence.* Think of genius as talent elevated to the *n*th degree. Or to turn it around, talent is genius on a human scale; and competence is wisdom on a human scale. Genius and talent are two points on the same curve of a cognitive trait. Think of wisdom as competence elevated to the *n*th degree. Wisdom and competence are two points on the same curve of another cognitive trait.

With this approach we undoubtedly take something away from both genius and wisdom. Something of these grand concepts will be lost in the analysis, but a measure of clarity will be introduced, amounting to a worthwhile trade-off. And by demystifying them we make them amenable to an examination, which is at least somewhat scientific and not entirely poetic.

Talent and expertise are also highly valued traits, but they are within the reach of most of us. Does it mean that many among us will attain either genius or wisdom? Not likely. But many of us possess talent and expertise (or competence)—traits approximating those two, if on a more humble scale.

Heeding Sternberg's cogent admonition, we will not aspire to a full understanding of either genius or wisdom; or of talent and competence, for that matter. We are concerned here primarily with their neurobiology, their cognitive and brain machinery. This is admittedly a limited perspective, leaving out ethical, social, and possibly other factors. But it is a crucial perspective, and one basically untapped.

To proceed further, we need to forge working definitions of *talent* and *competence*. Suppose we define *talent* through *novelty* and *creativity*. Talent is a particular ability to create, in one's own chosen field of endeavor, genuinely novel content that departs radically from the previously created body of work: novel ideas, novel art, novel technology, novel industrial products, novel social structures, and so on.

Suppose we define *competence* through the ability to relate the new to the old. Competence is a particular ability to recognize the similarities between seemingly new problems and previously solved problems. This, in turn, implies that a competent person has at his or her disposal a vast collection of mental representations, each capturing the essence of a wide range of specific situations and of the most effective actions associated with these situations.

The continuity between competence and wisdom has not escaped the attention of psychologists. According to Sternberg, a wise individual is perceived by others as someone endowed with a "unique ability to look at a problem or situation and solve it." Note that both the formal definitions of, and the commonsense intuitions about, competence and its supreme manifestation wisdom emphasize not only a deep insight into the nature of things, but also—and even more so—a keen understanding of what action needs to be taken to change them. The popular image is one of the people turning to a sage for guidance rather than for explanations. Both wisdom and competence are most valuable in their prescriptive powers. Keep this in mind for now; we will revisit the subject of prescriptive knowledge later.

Talent and its supreme form genius, and competence and its supreme form wisdom, exist both in unity and in contrast. They are two stages of the same life cycle. Talent is a promise. Competence is a realization. Genius (and talent) are usually associated with youth. Wisdom and competence are the fruit of maturity. Mozart's impish face is the face of genius. Tolstoy's gnarled face is the face of wisdom. The trade-off between wisdom and youth has been noted by philosophers, psychologists, and poets alike. Wisdom and competence are the rewards of aging.

While exceptions exist in either direction, both assertions are accurate at least in a broad statistical sense. In scientists, the age of groundbreaking discoveries peaks at thirty and then tapers off. Einstein the genius was the twenty-six-year-old who formulated the iconic discovery of the twentieth century, the special theory of relativity. Einstein the sage was the sixty-year-old who advised President Roosevelt on the matters of war, peace, and nuclear energy, the twentieth-century iconic menace.

In the creative journey of a genius blessed with a long life, it is often difficult to tell when genius ends and wisdom begins. The two are seamlessly blended together to propel the creative process of remarkable achievement well into the old age. While Michelangelo's greatest work, the ceiling of the Sistine Chapel, was completed when the artist was in his thirties, he directed the rebuilding of the Vatican's St. Peter's Cathedral and designed its great dome when he was well into his seventies.

Such seamless progression and blending of genius and wisdom is awe-inspiring, and it adds a finished quality, a satisfying culmination to a great life. But it is not always attainable. History is replete with examples of "unfinished genius" that failed to evolve into wisdom. Arguably, the short and violent lives of the great Renaissance painter Caravaggio and the rebellious French poet Arthur Rimbaud showed no discernable progression toward wisdom. Rimbaud's soul mate and lover, the great

Symbolist poet Paul Verlaine, managed, despite his scandalous excesses, a somewhat longer life of genius, but also died amid dissipation and debauchery, without the slightest trace of movement in the direction of wisdom. It has been said about the great Athenian general Themistocles that "he was greater in genius than in character." Likewise, it could be said about Caravaggio, Verlaine, Rimbaud, and probably Mozart that they were greater in genius than in wisdom.

By contrast, some individuals have relatively nondescript beginnings—in extreme cases are even dismissed as mediocre or worse—only to show indisputable wisdom later. This is often the case with political leaders. The Roman emperor Claudius, the first post-war chancellor of West Germany Konrad Adenauer, and the slain Egyptian president Anwar Sādāt can arguably serve as the cases in point.

More mundanely, we all know people with a "flash-in-the-pan" kind of unrealized brilliance, and we all know people who are somewhat ordinary yet supremely competent at what they do in their own quiet ways.

Wisdom and Problem-Solving

So genius and wisdom, and by extension talent and competence, do not always travel together, and in fact they often don't. Most people seem to recognize the difference between these highly desirable traits. Sternberg has studied how people from various walks of life perceive the relationship between creativity and wisdom. It turns out that most of his subjects viewed these traits as being positively but very weakly linked, and in some instances even as being negatively, inversely linked. Interestingly, the same study shows that both "wisdom" and "creativity" were viewed by the subjects as being better correlated with "intelligence" than with each other. This suggests to me that the very construct of "intelligence" is, in the minds of most people, an attempt to capture a sum total of many aspects of the mind, rather than a particular, distinctive aspect of the mind.

The belief that novelty-seeking is the attribute of youth and that wisdom is the attribute of old age seems to be shared by a lot of people. Psychologists J. Heckhausen, R. Dixon, and P. Baltes conducted a fascinating experiment in which they asked their subjects which human attributes appear at what age. Most subjects believed that curiosity and the ability to think clearly become dominant attributes for people in their twenties and that wisdom becomes a dominant attribute for people in their fifties. When asked to rank various attributes in terms of their desirability, wisdom was ranked among the most desirable traits. In a similar study, Marion Perlmutter and her colleagues found that most people associate wisdom with advanced age more than with anything else. This amounts to an interesting syllogism: If people believe that wisdom is the privilege of old age and also regard wisdom as one of the most desirable traits, then they also must believe that aging has its benefits, its positive side, and its unique and valuable assets.

In the minds of most people competence, like wisdom, is also the fruit of maturity. Understanding wisdom as an extreme degree of competence is consonant with the approach taken by psychologists Paul Baltes and Jacqui Smith, who define wisdom as "expert knowledge," a highly developed ability to deal with the "fundamental pragmatics of life" involving "important but uncertain matters of life." They place "rich factual knowledge" and "rich procedural knowledge" among the important prerequisites of wisdom and point out that the accumulation of such knowledge by definition requires a long life.

Following Sternberg's prudent (and wise!) admonition, I will refrain from discussing the concept of wisdom in all its richness. I will forgo the existential, self-actualizing, and moral aspects of wisdom, so cogently considered by Erikson, Jung, Kohut, and others. I will limit the scope of this book to one aspect of wisdom: the enhanced capacity for problem-solving. This admittedly narrow, morally agnostic approach allows a few villains into the book, along with many heroes. While realizing the limitations of this approach, I feel that it is a big enough slice of an infinitely rich concept to tackle in one book. Problem-solving is the one aspect of wisdom that we are most prepared to explore through neuroscience.

If wisdom and competence (or expertise) increase with age in all their aspects, then how does one reconcile this with the common assumption that one's mental powers decline with age? Or, to turn it around, if our memory and mental focus decline with age, then how is it possible that our wisdom and competence grow? What sets wisdom and competence apart from other manifestations of the mind and allows them to survive the ravages of aging?

5

PATTERN POWER

Kinds of Wisdom

What are the neural mechanisms that allow wisdom, competence, and expertise to withstand the detrimental effects of aging and, up to a point, of neurological illness on the brain? To begin our exploration of the subject, we need to further examine the concepts of pattern and pattern recognition, and of their role in our mental world. By "pattern recognition" we mean the organism's ability to recognize a new object or a new problem as a member of an already familiar class of objects or problems. The capacity for pattern recognition is fundamental to our mental world, as discussed briefly in the "day in the life of the brain" scenario. Without this ability, every object and every problem would be a totally de novo encounter and we would be unable to bring any of our prior experience to bear on how we deal with these objects or problems. The work by the Nobel laureate Herbert Simon and others has shown that pattern recognition is among the most powerful, perhaps the foremost mechanism of successful problem-solving.

The ability to recognize certain patterns is present very early in life, and other patterns are learned at much later stages. Most, and probably all, mammalian species have a relatively ready-to-use capacity for certain kinds of pattern recognition built into their brains. Does this mean that mammalian brains (including the human brain) contain "hard-wired" or "pre-wired" pattern-recognition devices? The answer to this question is probably less a matter of "yes or no" and more a matter of "to what extent."

Research has shown that even the most elementary pattern-recognition brain machinery requires some "finishing touches" to be provided by the environment in order to become fully functional. When such finishing touches (usually in the form of early environmental exposure to the appropriate sensory stimuli) are lacking, even the most basic pattern-recognition brain machinery fails to become fully operational. So most pattern-recognition processes are a blend of hereditary and environmental factors. But the relative contribution of nature and nurture varies from one kind of pattern recognition to another and involves vastly different time scales—from millions of years to merely years.

Certain kinds of pattern-recognition devices stored in our brains capture the "wisdom" reflecting the collective experience of all the mammalians over millions of years. Following the famous neuroscientist Joaquin Fuster, let's call this kind of wisdom "phyletic," or "wisdom of the phylum."[1]

This kind of wisdom was essential for the survival of so many species over millions of years that it is genetically encoded to a substantial degree. Or to be more precise and to avoid the teleological note that crept into the previous sentence, those species had a better chance of survival whose brains contained the "phyletic wisdom" in a relatively "ready-to-use" form. I am

[1]A stickler for the fine points of zoological taxonomy might refer to collective mammalian memory as "subphyletic" or "class." Phylum is a higher taxonomic entity. All the vertebrates are members of phylum Craniata, which in turn is divided into five classes: mammals, birds, reptiles, amphibians, and fish. I use the word "phylum" as a shorthand, because "class" has so many other connotations.

talking here about the emotional reactions that we all still have, such as the fear of snakes, the fear of precipices, the feeling of joy at seeing the sun at dawn, and the avoidance of fire. Research has shown, however, that even such basic reactions do not come completely formed and ready to use. They require some environmental exposure to the appropriate triggering stimuli at very early developmental stages.

Another example of such relatively ready-to-use (but still requiring environmental honing at early stages of development) pattern-recognition devices, or to use Fuster's expression "phyletic memory," are the neurons in the visual cortex tuned to respond to particular simple features in the environment. They fire when a line at a particular slope, an angle, or a contrast appears in the visual field. It is tempting to believe that the phyletic memory of the latter kind enables members of a particular species, or possibly of a whole cluster of species, to engage in sensory discriminations particularly critical for their survival. The world consists of a myriad of physical attributes engaging various senses, some of which we share with other species and some of which we don't (like ultraviolet vision or very high-frequency hearing). Not all of these attributes are equally important for various species—quite the contrary. Different species, or groups of species, depend for their survival on different kinds of information about the world they inhabit. So it stands to reason that they benefit from different repertoires of phyletic memories, and even different repertoires of sensory systems.

The Wisdom of Culture

Now consider a very different level of pattern-recognition devices: those crystallized in human culture. The word "wisdom" is not commonly used to characterize a group of people, let alone a whole species. But it can be, and by that reckoning we are a wise species. Each of us has at his or her disposal a rich assortment of patterns that are proffered to us on a silver platter called culture.

As we already know, the capacity for pattern formation and pattern recognition is not unique to humans. It is shared by every other species capable of learning. What sets us apart as humans is the powerful capacity for transmitting the repertoire of these patterns from individual to individual and from generation to generation through culture. In a rudimentary form this ability is present in the higher primates. Chimpanzees isolated from other members of their species are known to sometimes engage in unique behaviors suggesting nongenetic transmission of knowledge. Such behaviors are often interpreted as the evidence of rudimentary "culture." As a firm believer in evolutionary continuities, I tend to accept this interpretation. But even if we embrace the premise that they are worthy of the name, the primate "cultures" are inherently limited, since direct imitation is the only mechanism of knowledge transmission at their disposal. Without the availability of symbolic systems, its reach is rather humble.

In other species, pattern formation is an "everyone-for-himself" affair, and every member of the species is a cognitive Robinson Crusoe of sorts, in need of constructing his own mental world to model his island. Under this scenario, the pattern-formation potential is constrained by the computational power of a single individual brain and by the time span of a single individual life. With no or very little cumulative effect across individuals, this potential is rather modest. But we see the beginnings of the empowering effects of cultural, nongenetic knowledge transmission in great apes, chimpanzees, and gorillas, who are able to learn a rudimentary "sign language" taught them by their human handlers, yet are unable to construct such a language on their own. Cultural influence is able to exceed the generative power of an individual brain!

Unlike other species, we humans are spared the hardship of

discovering our world from scratch. Instead, we benefit from the incremental effect of knowledge accumulated gradually by society through millennia. This knowledge is stored and communicated through various cultural devices in symbolic form and is transmitted from generation to generation. Access to this knowledge automatically empowers the cognition of every individual member of human society by making it privy to society's cumulative, collective wisdom. If wisdom is defined as the availability of a rich repertoire of patterns enabling us to recognize new situations and new problems as familiar, then we truly are a wise species.

Much of what comprises human culture is a way of storing and communicating this collective wisdom from generation to generation. This allows each of us to come into the possession of wisdom, the discovery of which far exceeds the computational capacity of any single brain. This is a unique asset of human society and a powerful tool that has been crucial in propelling our success as a species. The cultural devices for knowledge transmission rely on a vast variety of symbolic systems, of which language is only one. But among them, language plays a special, hugely important role. It is a meta-device from which most other cultural devices flow. In addition to natural languages, we have at our disposal more specialized "languages," like mathematics or musical notation.

All these symbolic systems, languages, and quasi-languages are powerful tools of conveying specific information across time and space. We know about the city-states of Ancient Greece and their wars with the Persian Empire from Herodotus's treatises. We know about the Roman imperial conquests from Julius Caesar's *The Gallic Wars* and Josephus Flavius's *Judaic Wars*. And we know about the Chinese-Mongolian empire of Kublai Khan because Marco Polo wrote about it.[2]

---

[2]The latter fact is particularly interesting, and for a somewhat nefarious reason. If Marco Polo fabricated parts of his travel accounts, as has been suspected by

Language allows us to generate both true and false statements, as well as statements of unknown truth value. As it turns out, this generative latitude of language makes it an extremely adaptive and powerful device for modeling not only what *is*, but also what *will be*, what *could be*, and what *we want* and *do not want to be*.

Since language does not have built-in "generative truth filters" in a narrow sense, it becomes a particularly powerful tool of intentionality, extrapolation, and goal formation. The capacity for creating symbolic models not of the world as it is, but of the world as you want it to be, interplays with the so-called executive functions of the brain's frontal lobes to create truly goal-driven behavior. The emergence of the human ability to create mental models of the future, of the world as we want it to be and not merely as it is, probably represents the combined result of the development of the executive functions vested in the frontal lobes and of language.

Yet, language has certain built-in "truth filters" in a broad sense, and certain rules of languages model the natural laws governing our material world. We often reject certain statements as violating the rules of language not because they are unintelligible, but because their content violates some of the fundamental natural laws. For instance, the statement "I will go to the movies yesterday" is not unintelligible; it would be a perfectly legitimate statement in a world with a bidirectional flow of time, as the statement "I tripped and fell up" would be perfectly meaningful in a world with an opposite, or random, directionality of gravity.

Language is much more than the means of recording specific

---

some historians, then this is a classic example of language as a potent cultural device of propagating both information and misinformation, both true knowledge and false knowledge. The Ptolemeic texts positing that the sun revolves around the earth will have to be placed into the latter category (at least this is what we believe today). While in order to be a useful cultural tool the language has to model certain fundamental aspects of the world we inhabit, it does not have built-in "truth filters" for specific statements.

knowledge. Language also shapes our cognition by imposing certain patterns on the world. Without these patterns, the world around us would be an overwhelming kaleidoscope of disparate impressions. Each of us acquires a rich collection of patterns that represent the collective wisdom of society, and this spares us the hardship of discovering the crucial patterns de novo.

By learning the use and the meaning of words as children, we acquire more than a communication tool. We also acquire a taxonomy, a way of categorizing the virtual infinity of things, events, and impressions that *is* the world, and thus of making our world stable and manageable. Knowledge of word meaning is part of our system of patterns enabling us to recognize new things as members of familiar classes. By learning the lexical and conceptual structure of language, we acquire an understanding of complex hierarchical relations among things. And by learning the grammatical structure of language, we acquire the taxonomy of possible relations among things. No single lifetime would be long enough to work out all these categories and relations "from scratch." By coming into possession of this linguistic treasure trove, we come into possession of the knowledge and wisdom of generations. With the ever-improving ability to glean the intricate workings of the brain's microcircuitry, a time may come when we will be able to identify attractor-like phenomena in real biological brains, different attractors corresponding to different units of language: words, grammatical clauses, and such. Failure to utilize the "collective wisdom" implicit in language catastrophically cripples one's mental life. It has been long suspected, for instance, that the failure of language to exert its organizing influence on the senses plays a role in schizophrenia and contributes to the inner chaos of a schizophrenic mind.

Language embodies our collective experience of centuries and millennia and instructs us which distinctions are salient in the world and which are not. But by its very essence, wisdom is not merely declarative; it is prescriptive. The classic question addressed to a sage is less often "What is?" and more often "What shall we do?" Linguists have long been commenting on the predicative nature of language. Representation of *actions* associated with various things and attributes is central to the structure of language. Language as a pattern-recognition device enables us to do more than classify things. It enables us to decide how to act with respect to them.

Is language a "veridical" device? That is, does it contain the one and only "true" classification of the things around us? This would be a very difficult proposition to defend. Any large set of objects or attributes permits a proportionately large number of alternative classifications. The classification implicit in natural language reflects the attributes most salient for our cultures and species. Languages developed in the societies of talking dogs or talking dolphins (let alone talking ants or talking bacteria) would parse the same physical world in very different ways. Different wisdoms for different species indeed! So, above all, language is a pragmatic device.

The wisdom of the species inherent in language is neither genetic nor hardwired. Like its neural medium, the neocortex, language is a flexible device, readily capable of accommodating change. Unlike the phyletic memory, language condenses the wisdom of the species reflecting mere millennia of existence as opposed to millions of years, still very much a work in progress.

Needless to say, language involves a multitude of brain processes and cognitive operations. Among them, the closest to "hard wiring" is found in speech-sound production. It appears that infants are born with the ability to produce a vast array of speech sounds, and that this array is the same across all languages and cultures. With an immersion into a particular linguistic environment, a Darwinian process of sorts plays itself out: Certain articulations are reinforced, and others become lost. This is why an immersion into a linguistic environment before age twelve or thereabouts results in an accent-free command of language, and an immersion at a later age leaves one with an accent. So even at this most basic level of language development, a complex interaction of hereditary and environmental factors takes place.

Linguists often marvel at the extent of similarities across the hundreds of languages of the world. Some of them take this similarity as evidence of a very precise genetic determination of language, of the existence in the brain of a highly language-specific and language-dedicated, hardwired neural circuitry.

But I would argue that languages can be alike because in a very broad sense their users are alike, as are the environments they inhabit. To put it simply, the languages of the world are alike because we all are members of the same species, with similar biologies and needs, and we occupy similar ecological niches. We are all denizens of one world, not of many different worlds. The lexical contents of different languages are similar because their users are surrounded with similar things and engage in similar actions. And the grammars of different languages are similar because they reflect similar kinds of relations between things. But when certain environments represent a particularly drastic departure from the average circumstances, so do the languages of their inhabitants.

The example usually invoked to support the point is that Eskimo languages reportedly contain dozens of words for different shades of snow, which have no counterparts in other languages. The Khoisan and Hatsa click languages of the Southern African and Tanzanian tribes (which possibly reflect certain features of the ancient protolanguage of the first *Homo sapiens sapiens*) are believed by some linguists to reflect the adaptation to the peculiar acoustic features of the desert terrain. Likewise, the indigenous Guanches's whistling language (long since supplanted by Spanish) of La Gomera, one of the lesser known Canary Islands, reflects an adaptation to the peculiar local terrain and enables the islanders to communicate from one valley to the next. If through some strange mutation certain groups of humans were to become predominantly aquatic like dolphins or airborne like birds, their language would almost certainly be drastically different from other human languages. This is so because languages undergo cultural evolution shaped by their utility, both as tools of representation and as tools of communication, to the groups of people using them.

In his classic essay *The Sciences of the Artificial*, Herbert Simon makes a compelling argument that the complexity of an organism's behavior is to a great extent the reflection of the environment inhabited by the organism, and not solely of the organism's internal structure. In Simon's example the complex path traversed by an ant on complex terrain is more a consequence of the landscape, with its grooves, hills, and obstacles, than of the ant's nervous and locomotor systems. A small creature of an entirely different kind (a snail or a caterpillar, for instance) placed in the same environment will traverse roughly the same complex trajectory despite the fact that its own internal organization is very different from that of the ant's. This creature does not even have to be a living organism. A small robot placed in a similar environment will navigate a similar path. Likewise, our language is shaped less by the specifics of our neural organization and more by the specifics of the environment we humans share. This truly makes language the repository of the "wisdom of the species."

Another argument in support of genetically programmed hardwired "language instinct," to use Steven Pinker's memorable phrase, is the rapidity and ease with which children acquire language. It may seem implausible at first blush that the complex system of rules embodied in grammar can be learned with such astounding dispatch without being "built into" the brain. But recent studies in the field of complexity, notably Stephen Wolfram's work with "cellular automata," have shown that intricate organization may arise from simple rules with astounding speed, defying our common-sense intuitions by the rapidity with which it unfolds. Furthermore, a variety of other skills are acquired by children with equally astounding speed, which cannot be emulated by adult learners. Everyone knows that early training is necessary to become a really good musician, dancer, or athlete. On a more mundane level, a person who learned driving at the age of fifty is highly unlikely to match the driving skills of someone who started as a teenager. Rapid skill learning in youth and the partial loss of this ability in adulthood is not unique to language; it is a universal phenomenon, probably reflecting the time course of pruning, a phenomenon discussed earlier in the book. Does this mean that we have a genetically programmed hardwired "instinct" for every such skill? I don't think so.

I suspect that the very notion of the "language instinct" is the result of looking at the brain through an artificially narrow slit, of considering language in isolation from much of the rest of cognition, its mapping in the brain, its development and decline following brain damage. It is much more parsimonious and plausible to think that language is an emergent property made possible, once the neural circuitry in the brain reaches a certain level of complexity. According to this scenario, language does not rely on any specific, narrowly dedicated circuitry, but is a product of very complex but relatively general-purpose neuronal networks in the human brain.

This scenario is supported by current knowledge of the functional neuroanatomy of language, which arises with impressive consistency both from lesion studies and from functional neuroimaging studies. Today we know that, contrary to some earlier assumptions, language does not sit neatly in one particular "language-dedicated" part of the brain. Instead, various aspects of language are distributed throughout the neocortex by attaching themselves to different cortical regions, each in charge of representing certain aspects of physical reality: Cortical representation of action words is found near the motor cortex in charge of movements; cortical representation of object words is found near the visual cortex containing the mental representations of things; cortical representation of relational words is found near the somatosensory cortex containing the mental representations of space, and so on. This is precisely the kind of distributed picture that a self-organizing neural net, rather than a genetically programmed net, would come up with.

Am I saying that the internal structure of the brain has no impact on the nature of language and other symbolic systems at our disposal? That would indeed be a fallacy, particularly coming from a brain scientist! Of course the brain has an impact on these systems—a huge, crucial impact at that. But this impact is quantitative rather than qualitative. It sets the limits on the system's complexity rather than on its specific content. In an equally compelling insight, Simon suggested that the "wisdom bank size" is roughly the same for the "collective knowledge bank of the species" and for the individual knowledge bank. Both the number of words of natural language recognized by a literate human being (the repository of the "collective wisdom of the species") and the number of chess position patterns in a

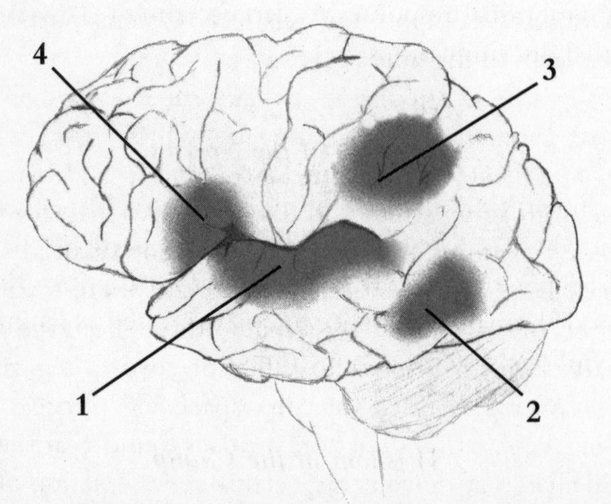

FIGURE 6. **Language Areas in the Brain.** *(1) Speech sound recognition; (2) cortical representation of object words; (3) cortical representation of relational words; (4) cortical representation of action words.*

grand master's memory (individual acumen in a particular field) are approximated by the same number: ~50,000. This figure should not be taken literally, but it may prove to be an interesting order-of-magnitude estimate of some important capacities of the brain for pattern formation, internalization, and storage, within a given domain.

So a "hierarchy of wisdoms" exists, each type of wisdom reflecting experience on vastly different time scales: millions of years for the phylum, thousands of years of civilization, and just years of your life. Each of them has its own mode of transmission:

Wisdom of the Phylum (or Subphylum, "Class")

This form of knowledge is expressed as a set of processes in the brain (to a substantial degree genetically encoded and transmitted), which are automatically triggered by certain stimuli or situations. This type of wisdom captures millions of years of experience of mammalian evolution and is expressed in humans as basic emotional responses to certain stimuli, as well as basic perceptual discriminations.

Wisdom of the Species

This form of wisdom is expressed as a culturally encoded and transmitted set of categories enabling us to parse the world in a species-adaptive way. This type of wisdom captures thousands of years of human experience and is expressed as language and other symbolic systems at our disposal.

Wisdom of the Group

This type of knowledge is the collection of skills and competencies possessed by a group of individuals with shared background (such as all the members of the same profession), which allows them to perform complex tasks, daunting to most people, in a relatively effortless fashion.

Wisdom of an Individual

This is the main subject of the book and we are well on our way in exploring it. But first we need to learn more about the cultural devices for pattern recognition, the foremost among them, language. It is commonly said that your language is as good as your intelligence. This is probably true to a large degree. But the opposite is also true: Your intelligence is as good as your language. As we have just learned, language is more than a communications tool; it is a rich repository of concepts, which allows you to pattern the world.

Russian Mavericks

Lev Semyonovich Vygotsky, a great Jewish-Russian psychologist, was the first to understand the importance of culture, especially language, in shaping individual cognition. He was a polymath by training, a maverick by temperament, and a uniquely colorful personality. His friend Aleksandr Romanovich Luria became his enthusiastic disciple and comrade-in-arms. In the 1920s, while they were very young (Vygotsky in his late twenties and Luria in his mid-twenties), they together began to sketch a uniquely original approach to psychology, which they called "historico-cultural psychology." The main idea of this approach is summarized in a mysterious sounding but profound premise that the cognitive operations of an individual develop, to a large extent, by way of "internalizing" various externally existing cultural devices. Based on their "historico-cultural psychology," Vygotsky and Luria went on to study how culture in general, and language in particular, shape individual cognition.

The "historico-cultural psychology" was first presented in a paper titled *The Tool and the Symbol*, an intellectual manifesto of sorts. Coauthored by Vygotsky and Luria in the late 1920s, it could not be published because it did not adhere to the increasingly oppressive prevailing dogma in the Soviet Union. The original Russian text was lost and only the English translation remained, prepared for a conference in the United States but never actually delivered. Forty years later, in the late sixties, the political climate thawed and their early ideas were exonerated. It was then that Luria discovered, to his dismay, the loss of the Russian original. Not one to be stymied by a challenge and always a practical man, he told me to translate *The Tool and the Symbol* from English "back" into Russian and make it sound like the original text. With a mixture of awe and amusement, I did just that, and our benign forgery was passed for the real thing. Today, it graces the opening volume of the collection of Vygotsky's writings, without an explanation of what had actually happened.

The "historico-cultural" approach to psychology forged by Vygotsky and Luria came increasingly under fire, and so was their crosscultural fieldwork with the tribes of Central Asia. The last straw came when Luria went to what is now Uzbekistan to conduct experiments with the native tribe members. The results of the study were fascinating. Optic illusions, commonly found among the members of modern Western society, could not be replicated with the Uzbek tribesmen. This suggested that even the most basic aspects of perception were under some degree of environmental and cultural control. Luria wired an exuberant telegram to Vygotsky, who had stayed behind in Moscow, consisting of four fateful words: "Natives have no illusions," followed by a row of exclamation marks. In the spirit of the times, the cable was intercepted and censored. In a society built on illusions, "having no illusions" could be easily construed as dangerous political blasphemy. Luria suddenly found himself in extremely hot water, denounced by the authorities as, among other things, a "Russian chauvinist," a surreally hypocritical accusation considering Luria's Jewish background and the tacit Russian chauvinism practiced by the Soviet empire itself. As a result of the incident, the crosscultural research was shut down and Luria was able to publish his Uzbek findings only four decades later, after the tentative political thaw of the Soviet Union in the 1960s and 1970s.

Meanwhile, Vygotsky and Luria found themselves increasingly under attack by the authorities, and the specter of arrest and deportation to a labor camp was looming increasingly ominous. As the 1930s unfolded, things were changing from bad to worse. For individual scientists, the threat of possible repercussions for political nonconformism ranged from public denunciation to murder.

Vygotsky's fate was dramatic and poignant. In 1934 he died at the age of thirty-seven and his ideas were suppressed in the Soviet Union for decades and were revived only many years later. His widow told me, years after his death, that she was convinced that his early death by tuberculosis was a blessing, because it had saved him from a far more tragic end; had he lived a year or two longer his life would have likely ended violently in the Gulag. Today, Vygotsky is regarded as one of the most seminal figures in twentieth-century psychology and cognitive science.

Aleksandr Luria, on the other hand, lived a long life and went on to become one of the world's best-known neuropsychologists. He was able to successfully navigate the Soviet political minefield and met with great, worldwide scientific acclaim and recognition in his own lifetime. He also became my mentor and my friend.

Luria would probably have never become a neuropsychologist had he inhabited a more benign environment. In the beginning of his career, the brain was relatively peripheral to Luria's agenda, and his first study of brain damage was designed to make a point that Luria himself later dismissed as naive and misguided: that the problem-solving skills of aphasic patients, deprived of the benefits of language, would deteriorate to the level of chimpanzees. This, of course, did not happen.

Luria's early interests and his early work concerned the relationship between culture and the mind, how the shared knowledge of society becomes the personal knowledge of the individual. Luria's early research was mostly developmental and crosscultural in nature, and he looked forward to a lifelong career in this field. But it was not to be. As the Soviet Union changed in the late twenties and early thirties, the exuberance of the first years of revolution gave way to the undisguised tyranny of the state, and authorities were increasingly applying Marxist doctrine to police every aspect of science. Among other things, this resulted in the denunciation of genetics and cybernetics as "bourgeois pseudoscience," and at the same time in the promotion of illiterate neo-Lamarckism in biology and agriculture.

In this climate, Luria's career took a very different direction. It was then that Luria, already a full professor of psychology at Moscow State University, went to medical school and subsequently began his association with the Burdenko Institute of Neurosurgery. This association was to last for almost forty years and provided Luria with the base for his groundbreaking work in neuropsychology. I have always suspected that Luria retreated into neuropsychology because it was less ideologically charged than other fields of psychology, and thus was relatively immune from the Party censorship.

World War II was Russia's great tragedy but also the country's only moment of relative glory under the Soviets. It was the only time in the seventy-three-year-long history of the Soviet regime that the interests of the state and the interests of the people were not at loggerheads when they intercepted in the collective effort to repel the Nazi invasion; it was the only event culminating in a victory as opposed to the string of colossal tragic failures that befell the country before and after the war.

For Luria the war provided both the purpose and the opportunity that tied him to neuropsychology for the rest of his life. He was charged with the task of developing neurorehabilitative methods for the wounded soldiers. In this capacity, he found himself surrounded with an abundance of penetrating gunshot wounds, which were to serve as the basis for his systematic investigation of brain–mind relations. This research culminated in two books that established him as the world's preeminent neuropsychologist: *Traumatic Aphasia* and *Higher Cortical Functions.*

Today, we are grateful that his complex life path took him to neuropsychology; without him neuropsychology would not be what it is today and, very possibly, simply would not be. Luria anticipated and in fact embodied, before virtually anyone else, the kind of fusion of psychology and brain science that we have witnessed over the last few decades under the name of "cognitive neuroscience." In Luria's time, and even a generation later, precious little interaction existed between the two disciplines. Even as recently as the seventies and eighties, a generation after Luria made his seminal contributions, academic psychology was dominated by people who were not only ignorant about the brain but proud of being ignorant. An infatuation existed with the thoroughly bogus notion that it is somehow possible to study cognition in its Platonic isolation, while leaving someone else to worry how it was "implemented" in the brain.

For their part, neuroscientists viewed psychology with patronizing disdain, and regarded complex behavior too mushy to be worthy of serious scientific examination. According to this view, in order to qualify as an object of rigorous scientific research you had to be a snail, or less. I recall using the term "cognitive neuroscience," then of very recent coinage, in a group of "mainstream" neuroscientists in the mid-1980s, and the disparaging looks I received suggesting that they thought the term to be a self-serving oxymoron. Luria was ahead of his generation in his ability to think about the brain and cognition with equal sophistication, and in his ability to integrate the two into a single narrative. In that he was truly a visionary. His *Higher Cortical Functions* was arguably the first monograph in cognitive neuroscience (albeit long before the term itself was born), an inauguration of a discipline.[3]

Today, the intellectual legacy of Vygotsky and Luria is pervasive, firmly embraced by the West and the East alike. It is no longer a solely Russian intellectual tradition, but rather a universal one, expanded and transformed in the process. Nor is Russia any longer the most fertile ground for their intellectual legacy to blossom. One can plausibly argue that today the most innovative continuation of Vygotskian and Lurian traditions takes place in North America and elsewhere in the West. In that sense, these traditions have shared the fate of another great Russian import, Stanislavsky's school of acting, which set firm roots in the United States in the form of Lee Strasberg's "Method."

---

[3]Not usually commented on, or even acknowledged, is the continuity between Sigmund Freud and Aleksandr Luria. As a very young man, Luria admired Freud and corresponded with him. Later, in the days of the worst Soviet excesses, when psychoanalysis was mocked and denounced from every official pulpit, Luria privately continued to speak about Freud with respect and with interest. Freud's early interest was in the brain and his early contributions were in the field known today as "behavioral neurology." Some of the most widely used terms of neurology and neuropsychology today were first introduced by Freud (like "agnosia"). Freud was one of the earliest proponents of the unity of the brain and the mind. But he felt that the science of his era was not ready for the "last frontier," not ready to tackle the mysteries of the brain. As a result, he focused on the mind, and psychoanalysis was born.

In the end of the twentieth century, when the scientific and general intellectual context was ripe, the fusion of neuroscience and cognitive science finally took place. One can think of Freud and Luria as two points on a long curve leading to this union. Luria's success in helping shepherd this fusion into being is to a large extent precisely due to his keen understanding of how the brain and culture interact, and the insights inherent in "cultural-historical psychology."

An Open-Minded Brain

In a larger scheme of things, the notion of the culturally molded mind, introduced by Vygotsky and Luria, leads to a very important corollary for our understanding of the biological machinery of the mind: *The brain comes pre-wired for certain kinds of pattern recognition but not for others.* This means that the brain must have some capacity, in fact huge capacity, to store information about various facts and rules, whose nature is not known in advance but is acquired by learning through personal experience or derived from culture. How can this be done?

Evolution solved the problem through the judicious application of the principle that "less is more." The "old" subcortical structures are preloaded with hardwired information representing the "wisdom of the phylum," and so are the cortical regions directly involved in processing sensory inputs: vision, hearing, touch. Motor cortex is also to a large degree "pre-wired."

But the more complex cortical regions, the so-called association cortex, have relatively little pre-wired knowledge. It has, instead, a great capacity to process any kind of information, to deal in an open-ended way with any curve ball the circumstances may throw at the organism. In a seemingly paradoxical way, the more advanced certain cortical regions are and the more recently they developed in evolution, the less "preloaded with software" they are. Instead, their processing power is accomplished increasingly by the ability to forge their own "software" as required by their survival needs in an increasingly complex and unpredictable outside world. This ability to forge "software" in the form of increasingly complex attractors is in turn accomplished by endowing these new brain regions with an open-ended capacity to deal with complexity of any nature. In contrast to the inborn, pre-wired processors, like the angle-specific neurons of the visual cortex, the pattern-recognition capability of these most advanced regions of the cortex is called "emergent," because it truly emerges in the brain, which is very complex but also very "open-minded."

This leads to a conclusion that is quite profound: The evolution of the brain is dominated by one grand theme, a gradual transition from a "hardwired" to an "open-ended-open-minded" design. As a result, the functional organization of the most advanced heteromodal association cortex does not resemble a quilt consisting of little regions each in charge of its own narrow function. To use the technical parlance of neuroscience, it is not *modular*. Rather, it is highly interactive and distributed. The heteromodal association cortex develops along the continuous distributions, called *gradients*, that emerge spontaneously, as dictated by brain geometry and neural network economy, and not by some preordained, genetically or otherwise, content-specific order. In the association cortex, functionally close aspects of cognition are represented in neuroanatomically close cortical regions. This congruence between cognitive metric and brain metric is exactly what one would expect as an "emergent property" in a self-organizing brain. I term this emergent principle of neocortical organization the *gradiental principle*. By contrast, attaining such congruence between cognitive metric and brain metric through genetic programming would have amounted to a tremendous, and unnecessary, waste of genetic information. Mercifully, this wasteful approach was rejected by evolution. Instead, evolution carved out in the brain design a space for a tabula rasa, but one powered by an exquisite neural capacity for processing complexity of any kind and filling itself with any content.

6

ADVENTURES ON MEMORY LANE

Memory Gauntlet

How does our brain, endowed with such powerful but open-ended abilities, acquire complex mental skills through individual experience and culture? What is the brain machinery of the "emergent properties" we have noted, including wisdom, competence, and expertise?

We will get to the matters of wisdom, but gradually. In order to navigate an uncharted territory—and the neurobiology of wisdom is such a territory—we must first link it to something better known and better understood: the adventures on memory lane.

One of the central points of this narrative is that wisdom is intricately connected with memory—a certain kind of memory, *generic memory*. Before we can tackle wisdom head on, we need to understand how this particular kind of memory works and how it is different from other kinds of memory. As we will see, a close and direct relationship exists between generic memories and patterns, and between the processes underlying their formation in the brain.

All, or at least most, memories are formed and stored in the brain's youngest and most elaborate part, the neocortex. In addition, certain memories require the support of various subcortical (or to be quite pedantic about it, *non-neocortical*) structures and other memories do not require such extra support. Those memories that depend on such additional structures are very vulnerable to decay and to the effects of neurological illness. By contrast, those memories that depend on the neocortex alone, and do not depend on the additional structures outside the neocortex, are relatively invulnerable to decay and can withstand the assault of neurological decline, even dementia, for much longer. Most of the memories of this latter kind are generic memories. But what is a generic memory? To understand this, we need to consider some basic facts of remembering and forgetting.

What did you have for dinner twenty-three years ago today? Don't worry. I am just trying to make a point: It is ridiculous to expect that anyone could remember such an inconsequential bit of trivia so many years later. Unless, of course, the dinner was a White House state dinner to which you had been invited. But had I asked you this question one day after the fact, you would have answered it precisely and unhesitatingly, state dinner or not. It *was* in your memory once, but now it is not; it's gone, forgotten. Memories for trivial, inconsequential events continue to decay very rapidly every hour following the events, and this decay is characterized by a steep power function. And thank God for that, because had you permanently kept all the memories that had ever, if fleetingly, been formed in your head, your head would be the mental equivalent of a city like Pompeii buried in lava and volcanic ash. Morsels of useful knowledge would be obscured by huge amounts of useless information—informational noise, informational trash.

There are some people with the uncanny propensity to remember everything without forgetting anything, although such cases are quite rare. Far from being a gift, this almost without exception proves to be a disabling, paralyzing curse. Aleksandr Luria described a case of a provincial newspaper reporter with the mixed blessing of clinging for the rest of his life to every memory ever formed, no matter how incidental and generally irrelevant. He described the unbearable and self-defeating condition of being constantly overwhelmed by a deluge of overlapping memories and images. Most of us are spared this fate, because what enters our long-term memory store is highly selective and most fleeting memories formed in our heads are not granted this privilege.

So forgetting as a normal phenomenon is, on balance, a good thing, as long as it is limited to inconsequential information. But forgetting may become abnormal, caused by various forms of brain damage, and then it is called *amnesia*. As we will see later, various forms of amnesia exist, as well as various degrees of its severity, ranging from relatively benign "senior moments" to a global catastrophic deficit when the patient loses the ability to remember what happened to him or to her ten minutes ago.

Amnesia may be caused by a number of brain diseases. They include traumatic brain injury sustained in car accidents or on the job, interruption of oxygen supply to the brain, viral, bacterial, or parasitic brain infections, diseases of brain vasculature, chronic alcohol abuse coupled with nutritional deficiencies leading to the so-called Korsakoff's syndrome, or severe seizure disorder, to name a few. These diverse disorders have certain things in common: They are likely to interfere with the brain's ability to form memories, to store them, and to access them when the need arises. We will revisit amnesia later; but for now let's focus on the ways normal memories are formed.

What do we mean when we say that certain knowledge has become part of the long-term memory store? A new memory begins forming the moment you encounter whatever it is you are learning: a new face, a new fact, or a new sound. The input engages the parts of your brain in charge of the senses, and then some higher-order brain systems in charge of analyzing and processing the new information and relating it to some previously acquired knowledge. This activity changes the very neural machinery engaged in the process, and the resultant change in the neural networks involved in receiving and processing the new information *is* memory. The process of memory formation has begun. New proteins are being synthesized, new synapses (contacts between nerve cells, neurons) are developing, and other synapses are being strengthened relative to the surrounding synapses. This is the essence of new memory formation.[1]

The first lesson to be drawn from this description is that memories are formed in the same brain structures, and involve the same neural networks, that participate in processing the information as it first arrives. In the past, many scientists believed that separate "memory warehouses" existed in the brain, removed from the brain regions originally involved in processing information that was being memorized. Today we know that no such separate "memory warehouses" exist, nor are there any neural "memory trains" shipping information from place A to place B. Instead new memories begin their neural life in the cortex and stay put right there for the duration of their "natural life."

In other words, the perception of a certain thing and the memory of that same thing share the same cortical territory; in fact they share the same neuronal networks. This was demonstrated with great elegance by Stephen Kosslyn. Using a high-tech research tool known as PET (positron-emission tomography), he identified the brain regions involved in mental imagery, the areas of the brain that were lighting up when the subjects were asked to bring before their "mind's eye" the images of various familiar things. The activated areas turned out to

[1]As we already know, neurons are not the only cells found in the brain; there are also glial cells. These cells, devoid of synaptic contacts, were thought until recently to have little if anything to do with information processing. It was assumed that the role of glial cells was limited to providing support and nourishment for the neurons. But today it is increasingly clear that some of them, particularly the glial cells called astrocytes, participate directly in neural computations by modulating the work of neurons.

be the same that were activated when the subject would actually see the objects.

Likewise, it was fashionable for many years to talk about "short-term memory systems" and "long-term memory systems," as if they resided in different parts of the brain. This misconception still persists in various professional and lay circles removed from state-of-the-art neuroscience. But in reality these are two stages of the same process involving the same brain structures, rather than two separate processes involving different brain structures.

Much in the brain's blueprint is plain impractical, defying the popular notion that the course of evolution is somehow inexorably and linearly directed toward improvement. For example, our brain stem contains a number of nuclei responsible for the brain's arousal and activation. They are packed so tightly in a single small area of the brain that damage to this area can in effect wipe out most of these nuclei with a single blow, producing a catastrophic impairment of arousal. This is precisely what happens in coma, which is caused by damage to this strategic area of the brain, the brain stem. A blueprint so devoid of redundancy and backup safety features would have been flunked in any school of engineering or design. A more "sensible" design guided by evolutionary wisdom, had there been such a thing, would have resulted in a much more distributed placement of the critical nuclei responsible for arousal and activation with ample backup and redundancy, so that not all of them end up in one neural "basket."

By contrast, the central feature of our memory machinery, the fact that memories are stored in the same networks that had received the information in the first place, would please any aficionado of design parsimony and economy, and anyone faithfully believing in the "wisdom of nature." When changes in the network become lasting and robust, the information becomes firmly ensconced in long-term storage. The changes that will have taken place in the network are chemical and structural. Synaptic contacts will have been altered and new receptors formed. The memory thus created will be robust and relatively invulnerable to any assault on the brain, whether it is traumatic brain injury, viral brain infection, or dementia.

Not So Fast!

These memory-forming changes in the brain do not happen instantaneously. They take time, usually a lot of time. They are painstakingly slow and they need a lot of help. In order for the memory to reach the stage of robust encoding, the process must be aided by certain other structures in the brain. Their role is to continue reactivating the critical neural networks in the neocortex, where chemical and structural changes gradually take place, even after the stimulus is long gone. Such processes of ongoing reactivation, also known as "reentry," are electrical in nature, involving loops of recurrent bioelectric activity in the brain. These loops can unfold on different scales and come in several varieties, which usually operate in concert. Some of these loops are far-flung, involving a number of distant regions, and these processes are called "reverberation" or "cyclic reentry." Donald Hebb, who foresaw so many mechanisms of neural computation, was the first to suggest that such loops play a role in memory.

Other loops are local, propagating right where the synaptic changes are taking place. The processes mediated by such local loops are called "long-term potentiation," or LTP for short. These processes have been the focus of more recent research. Two kinds of chemicals have been found to play a critical role in LTP: an excitatory neurotransmitter (chemical substance in charge of communication between neurons) called glutamate and its receptor, a molecule with an awe-inspiring name N-methyl-D-aspartate, or simply NMDA.

So the process of memory formation involves the interplay

between bioelectric, biochemical, and structural changes in the brain. To better understand the interplay between these processes, imagine walking down the street and noticing a useful telephone number on a billboard. You want to write it down but the street is just too busy and you have neither a pen nor a piece of paper on you. So on your way home you keep repeating the number to yourself and by so doing keep its mental representation alive, despite the fact that the billboard has long since disappeared from your visual field. You try to make sure that out of sight is not out of mind. But the process is full of hazards, and the mental representation you are trying to keep alive is fragile. Any street noise, any distraction, any passing thought may interrupt your mumbling the telephone number under your breath and the memory will evaporate. But with any luck you get home still repeating the number until you write it down in your notebook, and now the memory is finally safe.

The bioelectric reverberating loops in your head keep the memory alive much like the subvocal rehearsal of the number does, as you walk down the street: They ensure that the source of information is virtually present long after it had disappeared in actuality. Like the subvocal rehearsal, the reverberating loops are very fragile, unstable, subject to disruption by any number of physiological processes in the brain. It is a neurological obstacle course of sorts.

By contrast, once you have written the telephone number down, you have created a much more stable and robust structural record. It is several orders of magnitude more resistant to decay than the fragile loops we had just discussed. The structural memory trace may still perish. You may lose your notebook or it may burn in a fire, but the likelihood of this happening is relatively low. The formation of a memory in the form of a structural change in the brain is like writing the telephone number down. The memory has become much more robust, more invulnerable to any assault on the central nervous system, any effect of brain damage.

The propagation of the reverberating loops depends on a

number of brain structures found outside the neocortex. They include the hippocampi and surrounding structures, and the brain stem. The brain stem provides the general arousal level in the brain necessary for the reverberating loops to continue. The hippocampi do something more intricate and as yet not well understood. For now, let's assume that they make sure that the disparate cortical regions, where the engram is being stored, are coactivated together.

At the risk of annoying you, I will repeat an earlier point, because it is so important: *These structures are not the site of storage, the neocortex is.* But the hippocampi and other structures are extremely important for the formation of long-term memories, as long as the loops of reverberation must remain active.

These areas, particularly the hippocampi and the surrounding structures, are exceptionally vulnerable to the effects of dementia, and it has long been noticed that damage to these areas is likely to produce memory impairment. This is precisely what gave rise to the belief that the hippocampi are the seat of memories. But this leap to conclusion betrays a flawed logic. By this logic, the battery could be regarded as the seat of the information stored in your computer. But we know that it isn't; the hard drive is. Yet if the battery fails, you will lose the ability to store new information on the hard drive.

Once a memory has been firmly ensconced in long-term storage, the role of the hippocampi in maintaining it is drastically reduced. Presumably this happens because the cortical pathways between the far-flung components of the engram have become so well established that an external binding mechanism is no longer needed. But the processes of committing information to long-term store are full of hazards for the fledgling memory, a bit of a neural obstacle course, and they are excruciatingly slow. Just how slow these processes are, we came to appreciate only recently.

Earlier assumptions, based on animal studies, estimated the time course of permanent memory formation as a matter of hours and days. The experiments that led to this belief seemed to

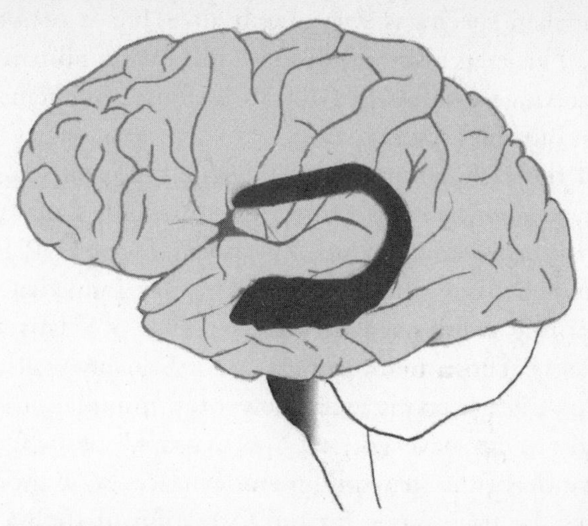

FIGURE 7. **Brain Regions in Charge of Memory.** *Necortical areas where memories are stored—light shading. Brain structures (hippocampi and brain stem) assisting in memory formation and recall—dark shading.*

be straightforward. A lab rat was taught a skill, usually navigating a maze. After the skill had been learned to a sufficient degree, the rat's head was "zapped" with electric shock. The assumption was that this indelicate procedure would disrupt the reverberating electric loops necessary for memory formation in the brain, and so would interfere with the memory processes still dependent on such loops. By contrast, the memories already committed to long-term, which have become part of the structural store, would no longer depend on the reverberating electric loops and would not be disrupted by the electric shock. With this in mind, the researchers manipulated the time delay between the skill learning and the application of electric shock, trying to determine the time lag within which the shock interfered with prior learning and beyond which it had no effect. This critical time lag in rats turned out to be of the hours-to-days duration.

But despite the favorite geneticists' saying, "Flies are flies but mice are human," mice are not human. Yes, the basic biology of all mammalian species is very much alike, but it is not totally identical. This tacit assumption that "mice were human" led to vastly inaccurate estimates of the time course of memory formation in our own species.

One of the early glimpses into the true time course of memory formation in humans came from the studies of "permastore." This term was introduced by psychologist H. P. Bahrick, who found that the rapid decay of memory immediately after initial learning is followed by a long period of relatively small further decay. Those memories that are reasonably well-retained three years after learning will show only minimal amount of further loss. They have entered "permastore," probably due to the formation of the structural memory trace. So it appears that in humans the time frame for the formation of such a trace is measured in years, not days, let alone hours.

Interestingly, the memories that end up in the "permastore" are not evenly distributed across the life span. The distribution of such memories is characterized by a "bump" corresponding to the ages of ten to thirty years old. It is possible that this period in human life is particularly rich in acquiring the most important knowledge, which serves as the basis of lifelong pattern-recognition abilities in the broadest sense.

But to set the record completely straight, it was necessary to study the effects of brain damage on human memory and to see which kinds of memories are lost, which kinds of memories are spared, and which kinds are first lost and then regained. A peculiar phenomenon, called "retrograde amnesia," proved to be particularly handy in shedding light on the time course of long-term memory formation in humans.

Amnesia Insights

Memory impairments, known in technical parlance as "amnesias," have always occupied a central place in neuropsychology.

Not surprisingly, a process as complex as memory can disintegrate in a vast variety of ways. Memory impairment is almost never global. It is almost always partial, producing a number of different forms of amnesia.

One of the main distinctions made in neuropsychology is between "anterograde amnesia" and "retrograde amnesia." Anterograde amnesia is a loss of the ability to learn new information after brain damage had set in. Retrograde amnesia is an inability to recall information acquired before the damage took place. Someone who suffered brain damage in a car accident last year and is now unable to recall what he read in the newspaper yesterday may suffer from anterograde amnesia. And if this person does not remember the name of the company that employed him for five years before the car accident, he is likely to suffer from retrograde amnesia. It is not uncommon to develop the two forms of amnesia together as a result of brain damage, and our poor fellow may be both unable to recall what he learned recently and to access the information acquired before the accident.[2]

The distinction between anterograde and retrograde amnesia depends on our knowledge of the exact time when brain damage occurs, which is not always easy to figure out. If a previously healthy individual suffered traumatic brain injury in a

[2]In a sloppy clinical parlance "retrograde amnesia" is often used interchangeably with "remote memory loss." In fact, these are different things. "Remote memory" is anchored to the present moment in time as the reference point. By contrast, "retrograde amnesia" is anchored to the time when brain damage is inflicted as the reference point. If someone suffered brain damage in a car accident ten years ago, then his poor memory for the events of nine years ago will be judged as "remote memory impairment" by any common-sense yardstick, yet it will not be retrograde amnesia. It will reflect the effect of anterograde amnesia (an inability to learn new information) already present nine years ago. By contrast, if someone suffered head injury yesterday afternoon and has no memories for the events of yesterday morning, this will be a bona fide evidence of retrograde amnesia even though yesterday morning is not particularly remote in time.

car accident, the exact time of the event is usually easy to establish. But in a case of dementia this is not possible, since in dementia the decline is gradual, unfolding over years. By the time a patient is diagnosed with dementia, he or she will have already been ill for a long period of time, often measured in years, not months.

Despite these diagnostic quagmires, the distinction between anterograde and retrograde amnesia has been very useful to neuropsychologists and neurologists for years. These two forms of amnesia often appear together; but for reasons idiosyncratic rather than logical, anterograde amnesia has always been given more attention, and it was assumed that it is more common and more severe than retrograde amnesia.

My own clinical experience contradicted this widely held assumption. I thought that we had witnessed the effects of a common logical error, the absence of evidence mistaken for the evidence of absence. (Because researchers did not pay nearly as much attention to retrograde amnesia as they did to anterograde amnesia, they did not find it.) In my own work I have always been particularly intrigued by retrograde amnesia, since I felt that it offered a unique window into the way knowledge is organized and stored in the brain.

Among other things, retrograde amnesia tells us about the time course of long-term memory formation. When memory of the past suffers following brain damage, not all memories suffer to an equal extent. Virtually without exception, relatively recent memories will be more affected than the memories for a very distant past. This phenomenon is known as the *temporal gradient of retrograde amnesia.*

Someone who suffered head injury in a terrible car accident may have lost his memories for the events of a month or two months, even a year or two years, before the accident, but he is more likely to remember the events of a decade or two decades earlier. The same is true for a patient suffering from dementia. This is why the common sense argument that someone's memory cannot be all bad, since he remembers the names of his grade school teachers, really proves nothing. A patient suffering from Alzheimer's disease will have such memories of very distant past preserved well into the advanced disease stages, while the memories for more recent events will be gone relatively early in the course of the disease.

The temporal gradient is counterintuitive. Many years ago I conducted a casual survey of a few friends from various walks of life, asking them to venture a guess as to what memories are more likely to suffer in brain disease: relatively recent or very distant. Guided by their common sense but unencumbered by the technical knowledge of brain science or neuropsychology, they all guessed wrongly without exception that the more distant memories would suffer first. This common sense–defying property of retrograde amnesia may serve as a great clinical tool for telling apart the memory loss caused by brain damage from the memory loss caused by psychological factors such as hysteria, or from plain malingering.

But the temporal gradient teaches us more than how to confound the unsuspecting public. It tells us volumes about how long-term memories are actually formed. Indeed, if memories remain vulnerable as long as they depend on the reverberating patterns of activation, then the extent of the temporal gradient allows us to estimate the amount of time necessary for long-term memory formation to run its course. And it turns out that retrograde amnesia may affect memories going back years and even decades.

It is known, for instance, that hippocampal ablation may result in retrograde amnesia going back as far as fifteen years. This means that it may take that long for a permanent, structural, relatively invulnerable long-term memory to be formed in the brain.

The process is gradual and incremental, rather than an abrupt, precipitous emergence of the long-term trace where there had been nothing before a second ago. The gradual nature of the long-term trace formation is illuminated by another peculiar feature of the temporal gradient—its "shrinkage." As we already know, it is not uncommon for a patient who had just suffered brain damage in an accident to have memory loss extending years and even decades back. But with the passage of time, some of the memories will return, and the recovery of memories will follow an orderly temporal course.

The span of the memory loss will "shrink." (This casual and somewhat inelegant word has in fact been adopted as a technical term in memory research, and scientists write about "shrinking retrograde amnesia," or "shrinking temporal gradient.") Like so many other features of retrograde amnesia, this mysterious process defies common sense. The shrinkage unfolds backward, memory for the more distant events returning before memory for the more recent ones. But the shrinkage is usually incomplete and the memories for the most recent events never recover. Just how extensive the permanent memory loss is varies from patient to patient and depends on the severity of brain injury. This permanent memory loss is genuine and intractable. No amount of hypnosis or "truth serum" will help recover the lost memories, and any attempt to do so will simply reflect a lack of neuropsychological sophistication.

The orderly and gradual process by which memories recover in "shrinking retrograde amnesia" tells us about the gradual nature of long-term memory formation. The farther along the process is, the more rapidly the memory will recover. But the memories whose formation has been assaulted by brain damage at very early stages are too fragile to rebound. They will be lost forever.

So the most critical obstacle on a memory's path into long-term storage is time itself. It takes years, or possibly even decades, for a long-term memory to form in the brain. Since there is no *perpetuum mobile* in the physical world, the reverberating loops stand a good chance of being extinguished on their own, and most of them are. Most reverberating loops become extinguished before the structural engram has had a chance to be formed. Nature appears to be very protective of the permanent memory store in the brain and the bar for being admitted into it is very high. So what kind of memories do receive preferential treatment in this arduous neural audition process? This will be discussed in the next chapter.

7

MEMORIES THAT DO NOT FADE

Generic Memories and Patterns

Enter generic memories, or "memories for patterns." Every new exposure to the same or similar thing in the environment—or, for that matter, to the same or similar information conveyed through language or by some other means—will breathe new life into the reverberating loop supporting the formation of memory about it, and will increase the memory's chance of making it into long-term storage. To use our billboard analogy, suppose you are walking home mumbling a useful telephone number you had noticed a few minutes ago. If you encounter another billboard with the same number along the way, the chances of your remembering it by the time you get home will be greatly enhanced.

The process is a bit Darwinian, as different memories compete for coveted but limited space in long-term storage. The more frequently encountered information usually wins, whereas the infrequently used information is likely to go by the wayside, into the dustbin of memory wannabes that never make it. One would think that the selection of memories for long-term storage should be determined by their importance, but we already know that there is no homunculus sitting inside the brain directing neural traffic. Even had there been one, he would have had a hard time predicting which information would prove to be important in the long run and which would not, since "importance" is mostly a prospective notion. Frequency of use becomes a surrogate "actuarial" marker of importance, since particularly pertinent information is likely to be invoked more frequently, and likewise frequently needed information is by definition important.

Nonetheless, importance may exert its influence on the formation of memories more directly as well. If in light of prior experience or genetically determined hard-wiring certain information is instantly recognized as "very important," then a brain structure called the amygdala becomes part of the memory-forming reverberating circuitry. This vastly facilitates and expedites the formation of a robust memory and gives it preferential treatment in the memory race. The Darwinian nature of many biological processes, including brain processes, has become increasingly apparent to neuroscientists over the last few decades, this being reflected in Gerald Edelman's memorable phrase "neural Darwinism." It seems that memory formation is no exception.

Different experiences activate different neuronal networks in the brain, and no two such networks are ever completely alike. But the closer and more similar the experiences are, the greater the overlap between the networks. The common core among the neuronal networks evoked by similar but not necessarily completely identical impressions ends up being activated particularly frequently and stands the best chance of quickly entering long-term storage.

This propensity of the shared properties of similar but not identical situations to be learned quickly is reflected in one of the most fundamental features of the learning process well known to psychologists: the phenomenon of overgeneralization. At early learning stages, both humans and animals tend to relate to similar but not identical situations as if they were indeed identical. The common aspects of the situations are learned much faster than the distinguishing aspects.

The shared network, found on the overlap of specific networks, will be the mental representation not of any single thing or event, but rather of the shared properties of a whole class of similar things or events. We have just traced the formation of a generic memory in the brain! Such generic memories are *memories for patterns*. The more generic a pattern is and the vaster the set of experiences on whose overlap it emerged, the more robust and invulnerable to the effects of brain damage it is. This means that abstract representations are generally better able to withstand the effects of brain decay than concrete representations corresponding to unique things.

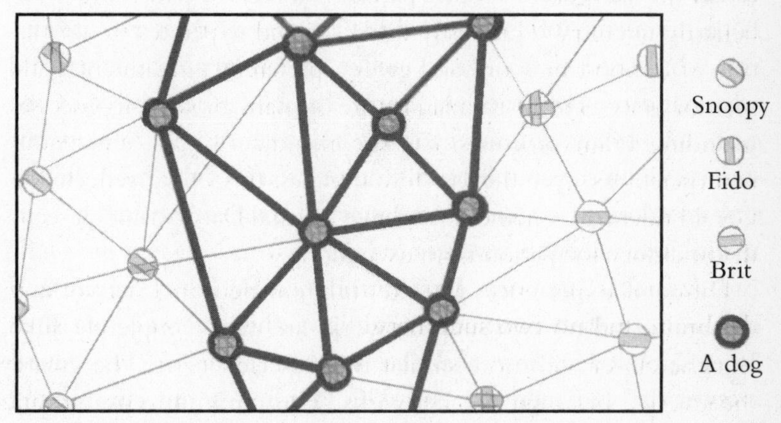

FIGURE 8. **Network Overlap.** *Specific networks—Snoopy the chocolate labrador; Fido the black doberman; Brit the fawn bullmastiff. Generic network—a dog.*

A typical pattern possesses a very interesting property. It contains information not only about the things you have already encountered, but also the information about things you may encounter in the future. This is so because a pattern captures the shared properties and features of every member of a whole class of things or events—all tomatoes, all chairs, all snowstorms, all political crises, all differential equations of a certain kind, all stock market crashes. Therefore, the pattern will help you deal with any member of the class that you may encounter in the future, by immediately informing you of all the essential properties of class members. The notion of *generic memory* or *pattern* can refer to the shared properties of entities of any kind, whether they are physical objects, social events, or verbal statements.

Now we understand why of all memories, generic memories or patterns are the most stable, the least vulnerable to any kind of neurological assault on the brain. This becomes particularly clear if one looks at the effects of retrograde amnesia. If different kinds of memories are affected differently in retrograde amnesia, then one could assume that these different kinds of memory are characterized by different degrees of robustness, due to the variability in the frequency of their use or the richness of their associations. Just what kind of memories suffer in retrograde amnesia, and what kind of memories are spared, has been the subject of inquiry and debate among neuropsychologists and neurologists for some time. In the course of this inquiry, some of the most important assertions of cognitive neuroscience have been made.

The first assertion involves the distinction between *procedural* and *declarative* memory. First introduced by Larry Squire and his associates, this distinction highlights the difference between "the knowledge *how* and the knowledge *that*." Procedural memory is the memory for skills. Riding a bicycle, playing tennis, and knowing how to tie a tie are all examples of procedural memory. By contrast, declarative memory is the memory for facts. Knowledge that there are seven days in a week, that Paris is the capital of France, or that World War II ended in 1945 are all examples of declarative memory. Like many distinctions in neuropsychology, the distinction between procedural and declarative memory is not absolutely cut-and-dried. How would you classify, for instance, the knowledge of chess or checkers moves? Is this truly knowledge, or are these, strictly speaking, skills? Despite this gray area, the distinction has been of great heuristic value to brain research. It has been commonly said that, with some exceptions, declarative memory usually suffers in retrograde amnesia, while procedural memory is usually spared.

Another influential distinction, first introduced by Endel Tulving, has been made between *episodic* and *semantic* memory. As we will see, this distinction further divides declarative memory into two more specific categories. Episodic memories are stored together with the memories of the context in which they were acquired. This may be the case both with momentous events or facts and with the most trivial ones. The knowledge that John F. Kennedy was assassinated in Dallas or the meaning of 9/11 is embedded in the minds of most people as the memories of personal circumstances surrounding these events. To put it simply, most people who lived through these events remember vividly where they were and what they were doing when the news reached them. The same is true for more mundane life events, like buying your first car or going on your first job interview: Not only are you likely to remember the make of the car or the name of the prospective employer, but also you will have actual memories of going through the motions.

By contrast, semantic memories are stored independently of the context in which they were acquired. Most people know that Rome is the capital of Italy, that Einstein was a great scientist, that there are seven days in a week, or that metal objects don't float in water, but they have no idea when and under what circumstances they had first learned these facts.

Like the procedural-declarative distinction, the semantic-episodic distinction also has its gray area. What is part of semantic memory for one person may be part of episodic memory for another, and vice versa. While 9/11 is part of episodic memory for most readers of this book, it will be an item of semantic memory for someone born long after the fact and learning about it from textbooks or from movies. By contrast, the knowledge that large bodies of water may have treacherous undertows is part of semantic knowledge for many people, but it is very much a part of my episodic memory store. This is due to the fact that I nearly drowned in the Mediterranean twice: both times many years ago, both times due to my own youthful recklessness, and both times very close calls, yet I was able to swim ashore to tell the story.

The episodic-semantic distinction has also been among the most influential in cognitive neuroscience and has been used in delineating the scope of retrograde amnesia. It has been commonly assumed that in retrograde amnesia episodic memory suffers and semantic memory is spared. But as it turns out, neither the procedural-declarative distinction nor the episodic-semantic distinction truly captures the fate of different types of memory in brain disease. It is not uncommon in biomedical research that established theories and opinions are challenged and eventually overturned by unusual clinical cases, which cannot be explained by these theories. My associates and I encountered such a case a number of years ago, and it changed our understanding of both the severity of retrograde amnesia and its scope. It is time to examine the memory problems of a fallen horseman.

Memories Lost, Found, and Spared

A victim of a horse-riding accident, "Steve" (a fictional name), suffered severe brain damage with memory loss and was admitted to the hospital where I worked at the time. Both anterograde and retrograde amnesia were present and both were extreme. Closely involved in his care, I approached Steve on numerous occasions in the course of the same day, and he had no memories of me, my name, or our prior encounters fifteen to thirty minutes later. This was an indication of severe anterograde amnesia.

Steve's retrograde amnesia was equally severe. He was a highly successful entrepreneur in his thirties, a caring husband and father. But post-accident Steve did not know any of that. He stated his age as seventeen. He gave his parents' address as his residence (he had indeed lived with his parent's at the age of seventeen). He denied having ever gone to college, being married, or having children. He could give a lucid account of his life events up to the age of seventeen and a somewhat patchy account of the events of the following two years. After that there was a complete blank covering seventeen years of his life, from the age of nineteen until his current age of thirty-six.

On the Richter scale of amnesias, where 0 is complete clarity of recall and 10 is complete memory loss, Steve's memory was at least an 8. But cases of comparable severity have been reported before, and we expected Steve's recovery to follow the course described in the standard neurology texts: rapid and substantial recovery from retrograde amnesia and somewhat slower and less complete recovery from anterograde amnesia. According to this rather common scenario, Steve would soon regain the memories of his past, but his ability to recall in the afternoon the main events learned from the *New York Times* in the morning would remain impaired. This was supposed to be the inviolate path of recovery from memory loss.

But as we were following Steve's recovery over time, we were witnessing, first with disbelief and then with fascination, the unfolding of a totally different picture. His ability to learn new information was steadily improving, so that only very subtle vestiges of anterograde amnesia remained. Steve was recovering his memory sufficiently to be able to regain the continuity of impressions from day to day and from week to week. Slight impairment of new learning was still apparent on formal testing, but for most practical, everyday purposes his memory was fine.

But Steve's memories for his past life, for his life before the accident, were not returning as expected. He continued to think of himself as seventeen to nineteen years old and revealed no knowledge of his life beyond that point. He had no memories of his college years or of his career as a successful entrepreneur. He knew his parents and his older brother but not his wife, children, or business associates. And there was not even the slightest suggestion of improvement in his ability to recall any of these things. Because Steve's ability to learn new information was improving by leaps and bounds, he was relearning a lot of facts about his life related to him by eager family members. But he made a very clear distinction between that which he truly remembered and that which he was told about his past life. This course of recovery, with anterograde amnesia receding and retrograde amnesia refusing to budge, was supposed to be a neurological impossibility. But here it was, and it changed my understanding of the mechanisms of memory and memory disorders.

As if this were not enough, Steve's memory impairment posed another puzzle. His retrograde amnesia was not limited to episodic memory; it clearly affected his semantic memory as well. This also contradicted the beliefs commonly held in the field at the time, which dictated that only episodic memory would suffer in retrograde amnesia. Steve did not remember his college years, which was an expression of his episodic memory loss. But neither did he remember that Madrid was the capital of Spain, that Newton was a physicist, or that Shakespeare wrote *King Lear*. This was very much a reflection of his semantic memory loss.

Steve's impairment of semantic memory was massive. Not only was it clearly present, but also in some sense, to the extent that this comparison could be made, his semantic memory was even more impaired than his episodic memory. Steve's episodic memory was intact up to the age of seventeen or thereabouts. But considering Steve's background one could safely assume that he must have learned these facts, for which he now had absolutely no recall, way before the age of seventeen. Shakespeare? Newton? Madrid? In upper-middle class, highly educated professional circles, to which Steve's family belonged, children generally learn this kind of stuff by the age of ten or twelve, if not earlier.

But how much of Steve's semantic memory was impaired? Was it a global loss or a partial one? As we continued to study

Steve's semantic memory, it was becoming increasingly clear that it was spared in some respects. He knew the number of weeks in a year, the color of tomatoes; he was able to give a pretty accurate estimate of an average male's and female's height and weight.

With my then research assistant Bob ("Chip") Bilder, I embarked on a more systematic study of Steve's memory. It soon became clear that while Steve's knowledge of *specific* facts was severely impaired, his knowledge of *generic* facts remained intact. His memory impairment was partial but persistent and full recovery failed to take place. Steve's case taught us that semantic memory may also be impaired following brain damage, but not in its entirety. The critical distinction seemed to be between generic and singular memory. Memory for specific facts was impaired, while memory for generic facts was spared. It appeared that of all the memories, generic memories don't fade.

Armed with the new insights provided by Steve's unusual (or so it seemed) memory impairment, my former graduate student Bill Barr and I embarked on more extensive research of retrograde amnesia. Once the old preconceptions were cast aside, Steve's profile of retrograde amnesia turned out to be the rule rather than an exception in various neurological conditions known to affect memory. Semantic memory for specific facts turned out to be severely impaired in traumatic brain injury, in Alzheimer's type dementia, and in Korsakoff's syndrome. But semantic memory for generic information was relatively spared in all of these conditions.

Generic Memories Don't Fade

The more we studied the profiles of memory impairment in various types of patients, the more important the distinction between generic memory and specific memory appeared to be.

Memory provides the content for our mental lives, but not all memories are equal. Some are much more resistant to the effects of any assault on the brain (and that includes aging) than others. The distinction between specific memories (describing unique things) and generic memories (describing the shared properties of whole classes of things) is so important because it shapes our understanding of the fates of different kinds of knowledge in brain disease and brain decay. Knowledge that Paris is the capital of France is an example of singular memory. There is only one Paris and one France, so this knowledge refers to a single entity. By contrast, knowledge that tomatoes are usually red is an example of generic memory, since there are millions of tomatoes on the face of the earth and this knowledge applies to all of them.

As a rule, generic memories are accessed much more fre-

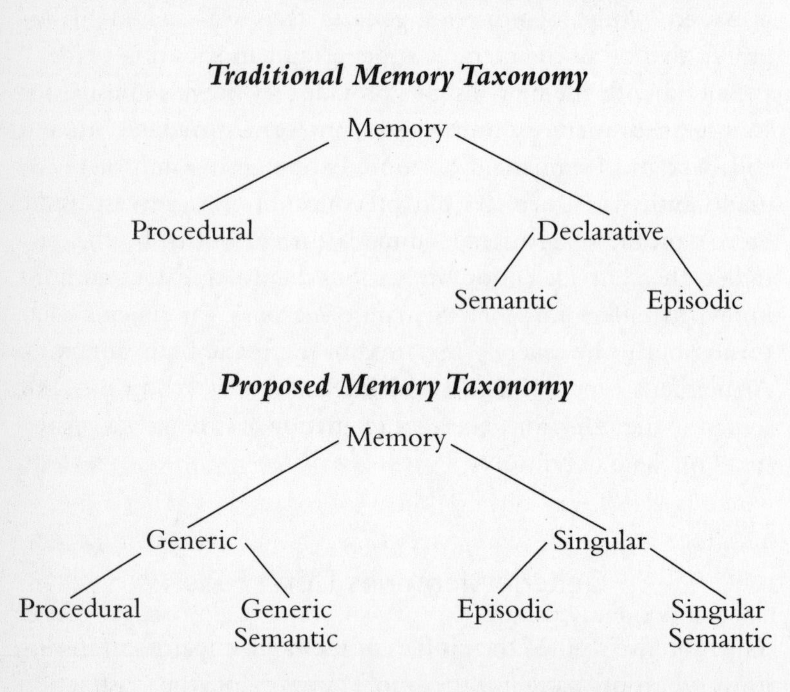

FIGURE 9. **How knowledge is organized.**

quently than specific memories. How often does an average American invoke the knowledge that Paris is the capital of France? A few times a month at most, whenever Paris is mentioned in the news, or when you plan your once-in-a-lifetime dream vacation. But you invoke the knowledge that tomatoes are usually red every time you walk down the supermarket's aisles or stick your fork into your daily lunch salad. Consequently, generic memories are much more robust than singular memories. Because of their high frequency of use, generic memories become committed to long-term storage more rapidly. As a result, they gain independence from the subcortical brain structures known to be particularly vulnerable in Alzheimer's disease and other dementias (to be precise, from the non-neocortical brain structures, since the hippocampi and the surrounding areas are also part of the cortex, just not the neocortex).

The relative invulnerability of generic memory becomes quite obvious if we consider two essential attributes of our mental life, which tend not to fade with age: language and higher-order perception. Although we tend not to think about these abilities as "memory," they are. In order to use language effectively, we need to "remember" which word refers to which thing, since the relationship is in most instances a matter of arbitrary convention and cannot be deduced logically. A language in which the word "chair" means table and the word "table" means chair would be every bit as effective as the language we use. And needless to say, the memory of the meaning of words, which is the basis of our linguistic competence, is generic memory, since any given word refers to a whole class of similar objects. A white Art Deco table, and a black-lacquered Chinese table, and a decrepit, rickety table in your neighborhood coffee shop are equal members of the same category and you refer to them with the same word, "table."

Likewise, our ability to recognize objects for what they are is also based on memory. Haven't you ever marveled at your own ability to come into contact with something you have never seen or heard before, and to instantly know what it is? You see an elaborately designed antique car on the street and you know that it is a car, despite the fact that you had never seen the likes of it. You hear a sound coming from outside, and you know that this is a dog barking, even though you had never heard a bark of this particular kind. To possess this ability, you must have a generic memory stored somewhere in your brain that captures the common characteristics of a whole class of things. You must have a previously formed pattern. Then, when you encounter an object containing enough of such shared characteristics, the generic memory will be evoked, and this is what object recognition is all about.

Thus, both language and higher perception are based on generic memories. Certain kinds of brain diseases may wipe these memories out, causing the patient to lose the use of words and the ability to recognize common objects. You may recall that in psychological and medical parlance these two types of symptoms are known as "anomia" and "associative agnosia." Such a breakdown of generic memories may be affected by stroke, traumatic brain injury, dementia, or some other brain disease. But the neocortex must take a direct hit for language or higher-order perception to suffer. Damage to the subcortical machinery alone will not affect them, since, as we now know, generic memories do not depend on this machinery. What's particularly important is that *language and higher-order perception are also resistant to the effects of normal aging*. This is so, at least in part, because they are independent of subcortical structures.[1]

---

[1]Linguistic knowledge and perceptual knowledge occupy such a special place in human cognition that psychologists don't even refer to them as "memory," even though in reality they are both examples of "generic memory." According to the prevailing terminological convention, "memory" as a technical term is reserved mostly for that which we call here "singular memory." By

An important point follows. Since singular memories depend on both the neocortical and subcortical brain structures, damage to either of the two, or to the connecting pathways, will cause their decay. This is a case of neurological double jeopardy. By contrast, generic memories depend on only the neocortex. This means that it takes a much more targeted kind of brain damage to affect them. While not totally protected from decay (nothing is), generic memories have fewer neurological Achilles's heels, fewer points of neural vulnerability.[2] This is why generic memories tend to not decay with age and may even be resistant to the effects of dementia up to a point.

The knowledge that frequent exposure to a particular kind of mental task speeds up the formation of a robust, long-term representation of the task and everything associated with it (including previous successful solutions) goes a long way toward understanding why certain kinds of memory are resistant to the effects of brain decay. But the formation of structural neocortical representation is not the only safeguard the brain develops to

---

contrast, large areas of "generic memory" (such as the knowledge of word meaning or the meaning of common objects) are excluded from the technical definition of "memory." Likewise, the term "amnesia" (memory loss) does not usually refer to the loss of memories for word meaning (when the patient knows what a common object is but cannot come up with its name); the term "anomia" is used instead. Nor does the term "amnesia" refer to the loss of memories for common object identity (when the patient does not recognize common objects for what they are); the term "agnosia" is used instead. For a clinician like myself, such a common-sense defying terminological jumble can be a source of perpetual confusion, since my patients, unencumbered by the knowledge of terminological intricacies and relying their symptom descriptions on good old common sense instead, constantly complain about "memory problems" when in fact they mean "anomia" or "agnosia."

[2]Steve's amnesia is a case in point. Without a known precedent, finding its cause took us some clinical detective work. It was caused by damage to the ventral mesencephalic section of Steve's brain stem, which disrupted singular memory but spared generic memory.

protect valuable information against the vagaries of neurological deterioration or illness. Other protection mechanisms are also at work.

The discovery of such mechanisms was made possible by state-of-the-art methods of functional neuroimaging. These methods, which include fMRI (functional magnetic resonance imaging), PET (position emission tomography), SPECT (single photon emission computerized tomography), MEG (magnetoencephalography), and others, made it possible for the first time in the history of science to observe the landscapes of physiological activation in a working brain of a living person, as the person is engaged in various mental activities. The introduction of these methods has changed the face of neuropsychology and cognitive neuroscience in a way not dissimilar to the one in which the invention of the telescope advanced astronomy. No field of inquiry can thrive on concepts alone, and the introduction of powerful new technologies (themselves products of novel ideas in other fields) usually plays a decisive role in scientific progress.

The application of these methods has led to the discovery of two additional protection mechanisms guarding frequently used knowledge represented in the neocortex. They are the mechanisms of *pattern expansion* and forging *effortless experts*. These two mechanisms work in concert.

In pattern expansion, with practice, experience, and repeated use the brain areas allocated to a particular motor, perceptual, and perhaps also cognitive skill expand and take over the adjacent parts of the cortical space. This was demonstrated in a variety of skill-learning experiments in the monkey by Michael Merzenich and his colleagues at the University of California, San Francisco. Even more to the point, similar effects have been demonstrated in humans. Alvaro Pascual-Leone has shown that in the blind, the cortical representation of the finger used for reading Braille is larger than the cortical representation of the same fingers in Braille-naïve seeing individuals. Likewise, the cortical representation of left-hand fingers is larger in string musicians than in other people. Such expansion makes the patterns more resistant to decay and to the effects of brain disease. To understand how this works, consider a simple Swiss-cheese model with a certain number of holes covering an area. If the number and size of the holes is kept constant, then the larger the total cheese-slice area, the larger the area spared by the holes will be.

While it may sound both irreverent and simplistic, the Swiss-cheese analogy is not that far off. In a number of age-related brain disorders, the brain is affected by tiny, discrete lesions, which destroy nerve cells and disrupt the communication between them. In Alzheimer's disease, the lesions are the infamous microscopic tangles and plaques, the debris of decaying and dying nerve tissue. In Lewy body disease, another primary degenerative disease, less prevalent and less well-known to the general public but every bit as malignant, the lesions are the microscopic Lewy bodies. In a different type of dementia, the so-called multiinfarct or small vessel disease, caused by a widespread disorder of brain vasculature, the lesions are tiny infarctions distributed throughout the brain. Whatever the etiology and pathogenesis of these lesions, they damage the brain tissue the way randomly thrown darts damage a bull's eye. But the greater the overall area of a bull's eye, the vaster will be its spared part—if not in proportionate then at least in absolute terms, which is probably what matters most for the preservation of a cognitive skill.

The pattern expansion mechanism is likely to be responsible, at least in part, for the puzzling phenomenon observed in the School Sisters of Notre Dame nun order in Minnesota, celebrated for its members' longevity and mental clarity in old age. The autopsies performed on some of the nuns' brains after their deaths showed clear signs of Alzheimer's disease, yet they enjoyed sound cognition with no indication of mental decline up to their last days. The nuns' brains were afflicted by Alzheimer's disease, but their minds were not. While the nun study is unique, I am reasonably certain that the phenomenon itself is not. The protection conferred by pattern expansion may account for the undiminished professional competence of many aging doctors, lawyers, and engineers who continue to function at high professional levels despite the occasional lapses of memory and attention in everyday life.

I like to refer to the other brain mechanism that offers protection of frequently used mental representations against decay as *effortless experts.* With practice and experience the metabolic demands of the neural tissue performing the task drop. This means that the brain can do a credible job of solving routine problems with fewer resources, including diminished blood supply. This discovery agrees well with the everyday-life observations of most of us. Tired, hungry, or under-slept, you will nonetheless be able to perform a familiar task, yet you will fail miserably at a novel task of equal or lesser intrinsic complexity.

Contemporary functional neuroimaging methods allow us to demonstrate this effect with great precision. In one of the earliest such studies, R.J. Haier and his colleagues used PET to study the brain glucose metabolism requirements in performing a novel complex task. The task used in the experiment was Tetris, the addictive spatial puzzle video game, which can get quite complex. They found that as the subjects were gaining mastery of the task, the metabolic demands were steadily dropping. After a few weeks of practice the metabolic demands of the brain dropped considerably, despite a seven-fold improvement in performance. Remarkably, the largest drop was seen in the subjects who gained the greatest proficiency with the task as a result of practice. This was truly the case of being able to do more with less!

Recent studies have demonstrated a similar "less is more" effect with fMRI in object-classification tasks. With increasing task familiarization, performance improves, while the task-associated cortical activation drops. In an ingenious experiment Ian Dobbins and his colleagues demonstrated that this effect is less due to the refinement of detailed problem analysis and is more due to the circumvention of such analysis altogether in favor of the automatic use of a learned response—very much a pattern-recognition type of shortcut mechanism.

The ability to perform a well-mastered task with fewer metabolic resources is a great source of protection against neurological assaults on the brain. Impairment of regional blood supply to the brain is relatively common in aging. The impairment may take a variety of forms, from mild to catastrophic, and it may affect different arteries and their branches. The most common mechanism behind such impairment is the narrowing of the blood vessels due to deposits of cholesterol and other types of debris along the blood vessel walls. As a result, the flow of blood, and consequently the oxygen supply to the region of the brain dependent on the clogged artery or small vessel, becomes reduced. A drastic reduction of regional blood supply may cause a stroke with subsequent irreversible tissue damage. But a subtle reduction of blood supply will merely dull one's cognition. The ability to perform complex mental tasks with diminished blood supply (and therefore with a diminished oxygen supply to the brain) serves as a powerful, albeit not infinitely powerful, protection against the detrimental effects of cerebrovascular disease on brain function.

Working together, pattern expansion and effortless experts increase the amount of brain space allocated to well-practiced cognitive tasks and decrease the metabolic requirements necessary for the effective performance of these tasks. While their power of protection works only up to a point, the combined effect of pattern expansion and effortless experts may be sufficient to counteract the effects of degenerative and vascular brain disease for a very long time, likely to be measured in years or even a decade or two.

As our "aerial view" of memory formation, of its gross neuroanatomy, is being increasingly refined, we are also beginning to understand the memory processes taking place on a much more microscopic scale. Exactly how long-term, permanent memories are formed in the brain is the subject of intensive research and much remains to be worked out. The cellular mechanisms of these processes are far from clear, and new information begins to pour in at such a rate that any book on the subject is likely to be somewhat out of date by the time it is published. Among the most interesting findings at the time of this writing is the possible role of prions in the cellular mechanisms of memory. These astoundingly sturdy proteins had been implicated until recently in only bad things, in incurable and catastrophic neurological disorders such as Creutzfeldt-Jakob disease, known also as spongiform encephalitis, and in mad cow disease. But the prions' astounding sturdiness, bordering on indestructibility, may yet turn out to be a useful ingredient in the formation of very stable memories.

The cellular mechanisms of memory are too complicated to be discussed in this book. It is clear, however, that memory-forming changes occur on the synapses, the tiny areas of contact between adjacent neurons. The changes may involve the growth of new dendrites, an increase in the amount of neurotransmitter (the chemical substance in charge of communication between the neurons), and an increase in the number of receptors, the molecules to which the neurotransmitters attach themselves. Any of these changes facilitate the connectivity within a group of neurons, so that activating any small subset of them will trigger a cascade of activation along particular paths. Compare this with the flow of water along previously formed grooves in the sand. Many scientists—and I am among them—believe that the formation of such facilitated neural paths *is* the formation of long-term memory, and their activation *is* an act of recall of previously stored information or an act of recognition of a particular thing as a member of a known category.

The grooves-in-the-sand analogy is useful, but only to a point. Every time you activate a previously formed memory, you change it ever so slightly by embedding it into a new context prompted by the unique circumstances of the mental activity at hand. As a result, the groove configuration will have changed ever so slightly as well. Just to make the point, I am thinking about a purple elephant with spiral tusks and a striped trunk. I am conjuring up this bizarre creature for the first time in my life as I am writing these words, basically putting on paper the first thing that comes to mind. In the process, I have activated my visual memory of the elephant, a well-established mental representation that I don't get to activate very often. But as a result of this frivolous exercise my mental representation of the elephant has become linked to my mental representation of "memory" as an abstract concept and to the mental representation of the grooves in the sand. Mechanistically, this means that the connections within the underlying neural networks have been reconfigured ever so slightly. This change may be ephemeral and not survive the rigors of neural Darwinism; but then again it may survive if having conjured up the example for the purposes of this book I will continue to use it in my future lectures to my students, thus resulting in a lasting change in the underlying neural networks. So memories undergo constant reconstruction and reconfiguration, as they are being called upon.

The changing, dynamic nature of neural nets may be the reason that some additional biochemical processes must take place to "reconsolidate" a previously formed memory once it had been used in the context of a new task. When these biochemical processes are interfered with, a memory that had resided in your head in a passive, stable form cannot be successfully "redeposited back" into this state after having been activated in a new context. It is no longer quite the same memory.

Brain Attractions

While the intricate details of how memories are formed and retrieved are still being sorted out, an increasingly elaborate insight into these processes comes from *computational neuroscience*. As with the rest of biology and psychology, brain science has traditionally been an empirical discipline, its general principles established through painstaking observation and experimentation. But ultimately the maturity of any discipline is judged by its ability to develop a theoretical arm.

Enter computational neuroscience. The choice of the adjective is, in my opinion, unfortunately bland, as it fails to convey the breadth and richness of this new discipline. I would have preferred "theoretical neuroscience" by analogy with "theoretical physics" and with similar connotations (even though I realize that the adjective "theoretical" in conjunction with "biology," let alone "psychology," has its baggage). In the past it often carried the connotations of winded verbiage and unprovable, speculative conjectures, an antithesis of the rigor and precision implicit in the term "theoretical physics." It may be for those reasons that people use the term "theoretical neuroscience" with a degree of caution and even reluctance, and feel safer with the austere connotations of "computational."

But today computational neuroscience is probably the most rigorous branch of brain research. At its inception, the methods nurtured in this new field involved mostly mathematical models of somewhat narrow, isolated processes in the brain. The advent of high-power computers has given rise to a peculiar hybrid of theoretical and experimental methods—computer modeling. The theory about the structure of a complex biological system is posited as a computer model, and then the "behavior" of the model is examined empirically, by making it perform various tasks and changing its various parameters. This blend of theory and experiment has yielded results far more powerful and unexpected than either of the two methods by itself. Some of these results bear directly on our understanding of the brain machinery of memory and they were obtained using the so-called *formal neural nets.*

Formal neural net modeling is among the most powerful and promising tools of computational neuroscience. Assembled from a large number of richly interconnected simple elements ("formal neurons"), they capture the most fundamental properties of the ways in which the actual biological brain works. Like in the real brain, a single element of the network, the neuron is limited in its abilities and cannot do much on its own. Like in the real brain, the problem-solving power of the network arises as a consequence of multiple interactions, both sequential and parallel, among neurons. The informational power of the network is everywhere and nowhere in particular. It is distributed throughout the whole net.

Any even moderately complex cognitive process unfolding in the real biological brain involves too huge a number of neurons and glial cells to permit an experimental analysis of all the important interactions among them. Simply put, the brain is a structure with too many moving parts and its most interesting properties arise from the multiple interactions among the parts, rather than from the parts themselves. But hidden as these multiple interactions may be from the tools of experimental research, many of them reveal themselves in dynamic neural-net models run on computers.

Faced with various tasks, formal neural nets exhibit amazingly brain-like properties. The most interesting among them is the rise of new abilities and skills that had not been explicitly programmed into the model by its designers. We refer to such new, spontaneously arising abilities as *emergent properties*. By acquiring such abilities on their own, neural networks in a sense truly "invent themselves." The networks exhibit these abilities when they have the benefit of explicit feedback about their prior success or failure (*supervised learning*), and even when no such feedback is available to them (*unsupervised learning*).

Among the most intriguing emergent properties are *attractors* and *attractor states*. An attractor is a network, a group of closely interconnected neurons with a stable pattern of activity in the absence of direct stimulation from the outside. These self-perpetuating patterns of activity are called "attractor states." The attractor states are possible because the connections between the neurons within the attractor are so strong (the grooves in the sand so deep, to use our earlier analogy) that the activation of any subset of neurons, even a relatively small one, is sufficient to keep the whole pattern going. This means that the same attractor will be activated in its entirety, as a whole, by activating any number of its various components. This property of attractors in the brain is sometimes referred to by the slightly dismissive term *degeneracy*, first introduced into neuroscience by Gerald Edelman. In fact, "degeneracy" is a fundamental mathematical property, extensively studied in algebra and symbolic logic. Degeneracy is also a very important property of biological attractors.

To better understand how attractors work, it may be helpful to be reminded of the original meaning of this term. The term *attractor* was borrowed by neuroscientists from mathematics. Originally introduced by the great nineteenth-century mathematician Jules-Henri Poincaré, it refers to a situation when an equation yields a single, constant solution for a whole range of numerical inputs. It was then said that such a solution "attracts" a whole range of specific numeric inputs into the equation. Another example of "attractor" can be found in Boolean algebra, where the same logical formula can be realized by a large number of input combinations.

Like a mathematical equation with attractor properties, an attractor neural net in the brain will be activated by a whole range of different inputs from the outside world, all activating the same net. While we recognize a short, black plastic pen as a pen, a long, red metal pen as a pen, and an ostentatiously rich gold pen as a pen, all of them produce very different sensory inputs.

Nonetheless, the same neural net will be activated by all three sets of inputs despite their differences, and this is how we recognize a pen as a pen as a pen.

To make matters even more intriguing, every attractor has a so-called basin of attraction, a set of similar activity patterns that tend to transform into attractor state. This means that a whole range of similar, but not identical, activation patterns are "recognized" by the system as being in some sense equivalent. The main attractor properties in the formal neural net, particularly the degeneracy properties, correspond to the propensity of a whole memory to be evoked by encountering any of its component parts. And an attractor with a basin state is like a generic memory, where a whole multitude of similar objects are recognized as members of the same category.

Even though the ideas of attractors and attractor basins come from computational models, the possibility that they capture the essential features of real memory formation in the biological brain is tantalizingly seductive. John Hopfield, one of the pioneers of neural net modeling, was among the first to propose that attractors are in fact memories.

At the very least we know that attractor-like circuits exist in the brain. Their function is not entirely clear, but the evidence in support of the "memories are attractors" hypothesis continues to accumulate. Some of this evidence comes from "morphing" experiments. Most of us have seen Michael Jackson's music video "Black or White," where this technique first appeared, with morphing faces: female faces morphing into male faces, old faces morphing into young faces, Asian faces morphing into Caucasian faces. The same idea has been applied to neuroscience experiments. With the use of computer graphics one can create a continuum of images morphing one animal into another: a dog into a cat or a cow into a camel. Suppose you ask human subjects to classify the computer-generated creatures corresponding to various points on this continuum into two groupings corresponding to the two original animals. You can do the same with computer-synthesized or blended voices uttering sounds of language by morphing vowels: "A" into "O," "O" into "U," and so on.

The classifications of such computer-generated items by human subjects usually produce amazingly discrete boundaries: up to a certain point on the morphing continuum all the items are assigned, unhesitatingly and consistently, to one category; beyond that point they are assigned to the other category, equally unhesitatingly and consistently. The discreteness of these classifications is exactly what one would expect in a brain with distinct attractors and a distinct basin space linked to each attractor.

Another pioneer of neural net modeling, Stephen Grossberg, developed a powerful *adaptive resonance theory*, or ART. According to the ART model, recognition and "making sense" of an external event takes place when the sensory input from this event to the brain "resonates with," or matches, one of the previously formed networks, attractors. According to this model, the act of recognition is nothing other than the reactivation of a previously formed neural net. This device is being increasingly accepted by neuroscientists as the model of what actually happens in the real human brain when we recognize an object or retrieve something from memory.

The brain aficionados among the readers of this book might be curious about the relationship between an attractor and a *module*. The term *module* was popular in cognitive science in the 1980s and 1990s and still remains popular in some circles. It implies a structurally compact, circumscribed, and "informationally encapsulated" unit in the brain, dedicated to a very specific mental operation, sometimes a rather complex one. As alluded to previously, communication between different modules was presumed to be extremely limited with virtually no overlap in either their function or their circuitry. It was fashionable for a number of years to regard such "modules" as the basic building blocks of cognition and of the brain. The modular view of the brain was a peculiar resuscitation of the nineteenth-century phrenology, retouched and dressed up as a state-of-the-art innovation.

To me the notion of modularity of higher cognition was the intellectual equivalent of the Visigoth invasion trampling over more nuanced understanding of how the brain works, and I am on record saying so. I fought it tooth and nail, usually in the minority, sometimes in a minority of one, publishing journal articles with irate titles like "The Rise and Fall of Modular Orthodoxy," anticipating the demise of the "module muddle." The demise was not very slow in coming. Today, modularity has been thoroughly debunked, rejected, and all but completely discarded by much of the neuroscientific community. The cognitive module is sometimes referred to, sarcastically, as the "grandmother cell," a neuron where the image of your grandmother is stored. Don't look for it. It exists only in the heads of die-hard devotees of the outdated modular theories. Unless you are one of them, there are no "grandmother cells" in *your* head!

But is an attractor really a module in disguise, a "grandmother cell" by another name? Did we invent a high-tech computer-simulation term to merely rename a conceptual mummy of the past? The answer to that is a resounding "No." A module is presumed to be innate. An attractor is emergent. A module is supposed to be functionally encapsulated. Numerous attractors share the same neural components. A module is supposed to be structurally encapsulated. An attractor can be, and probably more often than not is, distributed across a vast territory of cortical areas. The latter point is illustrated by a common observation. Suppose you are trying to come up with someone's name in a conversation. It is on the tip of your tongue but keeps eluding you—until the person in question enters the room. The moment you see the grinning face, the name immediately comes to mind. The epiphany-like recall takes place despite the fact that the person does not wear a name tag—nor, needless to say, is the person's name written on his or her forehead.

For this name-recall phenomenon to take place, a network

must have existed in your head that incorporates both a visual component containing facial information and an auditory component containing name information. Despite the fact that these two kinds of information inhabit very different cortical areas (the parietal lobe for facial information and the temporal lobe for name information), they are intertwined into a single attractor. And the whole attractor is activated in its entirety as soon as even a small subset of its component-neurons is activated.

This, in a nutshell, is the mechanism of generic memory. Just how powerful a cognitive tool generic memories are will become clear in the next chapters.

8

MEMORIES, PATTERNS, AND THE MACHINERY OF WISDOM

The Virtues of Mental Economy

Feats of wisdom (or on a humbler scale, displays of expertise) usually strike an awed observer as a near-instantaneous, seemingly effortless "knowing" of the solution to a seemingly thorny, unexpected problem. Wisdom is also the ability to anticipate the events that catch most people completely unaware. We have already established that the phenomenon of wisdom, with all its complexity, cannot merely be reduced to the capacity for high-level pattern recognition. But we have also established that such pattern-recognition capacity comprises a very important element of wisdom, which implies that a person endowed with wisdom has the ability to recognize an unusually large number of patterns, each encompassing a whole class of important situations. As we already know, this ability is the result of a large number of attractors stored in one's brain. It takes time for the pattern-recognizing attractors to accumulate and form. The patterns that enable us to find quick solutions to a wide range of problems are generic memories. The arsenal of these generic memories accumulates with age.

Also accumulated with age is the facility for intuitive decision-making. Intuition is often understood as an antithesis to analytic decision-making, as something inherently nonanalytic or preanalytic. But in reality, intuition is the condensation of vast prior analytic experience; it is analysis compressed and crystallized. In effect, then, intuitive decision-making is postanalytic, rather than preanalytic or nonanalytic. It is the product of analytic processes being condensed to such a degree that its internal structure may elude even the person benefiting from it. The "postanalytic" nature of intuitive decision-making was pointed out by Herbert Simon.

The advantages of such mental condensations were "discovered" by evolution millions of years ago and have been utilized across generations of various species. Certain agents in the environment, like a snake, are "recognized" as danger through an instantaneous, automatic, extremely efficient process not requiring any deliberation. One can think of this mechanism as a form of "phyletic" wisdom, a notion introduced by Joaquin Fuster and discussed in an earlier chapter. Like every highly generic mechanism, phyletic wisdom is statistical in nature. It works to our advantage most of the time, maybe the overwhelming majority of the time, but not always. And it operates with the near-absolute force of a hardwired mechanism, which it is. The seat of this hardwired, automatic response mechanism is in the amygdala, a small collection of nerve cells found on the inside of each temporal lobe.

I came to fully appreciate the force of such hardwired decision-making condensations on a trip to Kenya many years ago. Among many other things that tourists do, I visited a crocodile farm, where a just-hatched baby crocodile was proffered to me. The tiny creature was barely the length of my palm, skinny and obviously harmless. Yet as I extended my arm to touch the creature (a conscious process directed by the neocortex), an unfathomable force pulled my arm in the opposite direction away from it (an automatic process directed by the amygdala). I was witnessing this neural tug-of-war with utter disbelief and with an odd feeling of being a passive observer of the inner workings of my own brain, rather than their empowered agent. To my amazement, the amygdala prevailed and I found myself unable to touch the baby crocodile. From a rational standpoint, the situation was utterly ridiculous, but the hardwired mechanism, honed through generations of species, had the last word. A similar reaction to snakes has been reported by a number of people; and I must admit to a bit of a shiver every time I see a large snake suspended from the shoulder of a street entertainer, a scene not uncommon in many urban environments. The thought of approaching the creature and touching it never crosses my mind, even remotely.

Just as the amygdala contains neural condensations embodying the phyletic wisdom that developed over millions of years, the neocortex contains neural condensations that embody individual wisdom (or competence) developed through the lifespan. These condensations come in the neural forms of the attractors that we discussed before. Just as in the case of my baby crocodile, filtering information about the world through such cognitive templates may occasionally misfire.[1] But on the whole they are extremely adaptive.

The intuitive decision-making of an expert bypasses orderly, logical steps precisely because it is a condensation of extensive

[1]The Australian neuroscientist Allan Snyder believes that the price we pay for relying excessively on rapid pattern-recognition mechanisms may be more considerable than we think. To find out just what the cost of dependence on these mechanisms is to our mental life, Snyder has embarked on a study that temporarily disables the pattern-recognition brain circuitry by passing a weak magnetic signal through his subjects' brains (the technique is known as transcranial magnetic stimulation or TMS). He claims that as a result of this manipulation, his subjects temporarily gain mental skills not previously available to them. In particular, their drawing ability is reportedly much enhanced and becomes richer with detail; and so, too, may be their other abilities. Snyder is a terrific scientist and a great friend, I declined an invitation to be a subject in his TMS study. And so we will never know how this book, which I was writing at the time of my visit to Snyder's lab at the University of Sydney, may have improved had I accepted.

use of such orderly logical steps in the past. It is the luxury of mental economy conferred by vast prior experience. The great physicist Richard Feynman was reportedly able to scan several pages densely covered with arcane mathematical formulas and casually conclude: "Looks about right." Effortlessly postanalytic!

A simple, everyday illustration of mental economy conferred by previously accumulated knowledge is found in our ability to read the newspaper without, strictly speaking, reading it. I open an issue of a major newspaper circa the end of year 2003 and skim through the headlines: *Milošević Ill Again . . . Schwarzenegger Gaining . . . Bali Bomber Sentenced. . . .* I do not need to read the whole articles to know their content. My previously accumulated knowledge about current events allows me to infer the content with such great accuracy that if I were to actually read the reports word for word, I would not have learned much more than I had already inferred. The war-crimes trial of the former Yugoslavian president Slobodan Milošević has been put on hold again because of his claim of poor health. The bodybuilder-turned-politician is ahead in the California gubernatorial race. The trial of the Muslim fundamentalist who blew up a discotheque in Bali is finally reaching a conclusion.

I would have been able to extract all this information without the prior knowledge too. But then I would have had to read the news reports very closely. I would have spent at least thirty minutes, maybe an hour, extracting the information from the text, and the process would have taxed my attention, memory, and linguistic abilities. But with the benefit of prior knowledge, the whole process boiled down to near-instantaneous recognition, was marvelously effortless, and took all of thirty seconds. Here is mental economy for you! Of course, my newspaper example is a far cry from decision-making in demanding, complex situations. But the principle of previously accumulated patterns serving as the mechanism of mental economy operates in a fundamentally similar fashion across various, seemingly very different situations.

The neural benefits conferred by such mental economy are considerable, and their value to the individual increases with age. To understand why this is so, the notions of "mental reserve" or "mental resources" are often invoked, and it is assumed that they tend to dwindle with age. These concepts have gained prominence among neuroscientists concerned with cognitive aging, despite their somewhat mysterious ring. This reflects an attempt to capture some elusive aspects of the mind, which in the lay parlance are referred to as "mental energy" or "clarity of thought." Frankly, I have always felt that "mental resources" is one of those terms that creates the illusion of understanding through the invention of a new name for an old quagmire. (There are plenty of those in science!)

We don't know exactly what determines the amount of "mental resources" in a given individual. By way of sheer speculation, it could be the amount of oxygen made available to the brain through the bloodstream, the density of neuronal connectivity, the speed of electric signal transmission along the axon, the concentration of critical neurotransmitters in the synapse, or the combination of all of the above. Whatever is behind it, the amount of "mental resources" varies from person to person. But the mental economy, made possible by the mechanism of pattern recognition, enables a person to solve very complex mental tasks with a minimum expenditure of mental resources. Effectively, mental economy inherent in pattern recognition counteracts the decline in mental resources presumed to take place in most people as they age.

A lazy, untrained, and "unpatterned" mind is sometimes seduced by the apparent ease and effortless nature of "postanalytic" decision-making and is tempted to emulate it. Far from being postanalytic, such a pathetic display will most assuredly be "fake analytic." A recently fashionable educational trend teaching grade school and high school mathematics through impressionistic quantitative "estimations" rather than explicit computations is the worst example of such a cognitive fake.

So generic memories are pattern-recognition devices. The more we examine their power in cognition, the more impressed we are with Herbert Simon's early insight that pattern recognition is the most common and efficient problem-solving device at our disposal. Does this mean that every pattern qualifies as the element of wisdom or even of as an element of competence? Probably not, lest we trivialize these concepts unduly. But the more numerous and generic such patterns are and the greater the extent to which they facilitate an effortless and instantaneous solution of a wide range of important problems, the more such patterns qualify as the elements of wisdom. The more generic certain patterns are, the more redundant are their neural representations and the more resistant they are to the effects of brain deterioration and dementia. The more frequently such patterns are activated in the course of mental activity, the more invulnerable they are to the effects of cognitive decline. The repertoire of patterns grows with age. So aging is the price we must pay for accumulating wisdom patterns.

Earlier we have examined the relationship between competence and wisdom. How meaningful is this distinction? Our culture is dominated by the penchant for finite taxonomies, stark dichotomies, and binary distinctions. But reality is more often continuous and graduated than demarcated with distinct boundaries. I recall the incessant arguments I had as an adolescent with my equally precocious cousin who was roughly the same age. The arguments were about greatness and where to place the threshold defining it. We both agreed that Beethoven was a great composer, Rembrandt a great painter, and Tolstoy a great writer. But how about Béla Bartók, Francisco Goya, or Theodore Dreiser? Were they also great, or "merely" outstanding? In retrospect, of course, the debate was naive and fundamentally futile. The writings of scholars like Harold Bloom and Charles Murray notwithstanding, no clear-cut self-evident boundary exists between greatness and "outstandingness," and none exists between wisdom and competence. They are a matter of degree, subjectivity, and value judgment.

"Bundles of Habits"

As we established earlier, wisdom and competence come with age. Does this mean that as we age, we acquire these precious traits as a matter of course, the way we acquire gray hair and wrinkled skin? (That would be nice, wouldn't it?) But it does not happen in an assured, preordained way. In a book of interviews published by the renowned Australian radio journalist Peter Thompson, the subtitle is as important as the title: *Wisdom: The Hard-Won Gift*. The gift of wisdom is a reward, not an entitlement. It has to be earned. And likewise you have to work for your competence.

To revert to the language of the brain, both wisdom and competence are attained through the accumulation of attractors allowing pattern recognition in important situations. Well, then, it stands to reason that some people spend a lifetime accumulating such patterns, and others . . . less so. Every human being accumulates a certain pattern-recognition capability in the course of his or her lifetime. But not every human being accumulates the patterns necessary for the solution of problems of genuine importance to a significant number of other people. Generally speaking, people who have spent their lifetime dealing with strenuous mental challenges and who have been good at it, in other words people who are both bright and have been mentally active most of their lives, are rewarded with extra mental resistance to the effects of aging.

This became quite apparent when the relationship between reasoning ability and general knowledge (including language vocabulary) was examined. In people with low reasoning ability, general knowledge and vocabulary were either constant as they aged or showed actual decline. But in people with high reasoning ability both knowledge and vocabulary actually continued to increase with age—all the way to the age of eighty years old!

So it appears that the gift of effortless and powerful pattern recognition as a way of solving problems that faze other people is the culmination of and the reward for a lifetime of facing up to such mental challenges. In those who earned this reward, the gift of wisdom, to use Peter Thompson's turn of phrase, has an amazing staying power in the face of aging and of all manner of neurological assaults on the brain. The great American psychologist William James was right when he said: "Could the youth but realize how soon they will become mere walking bundles of habits, they would give more heed to their conduct while in their plastic state."

Those in whom "the bundles of [acquired] habits" include genuine competence continue to reap its benefits well into old age. Today, an increasing number of aged individuals elect to remain active in the workplace. This is a very welcome and demographically realistic development. But it has also triggered the concerns that their performance on the job would be compromised because of age. But the concerns proved to be basically unfounded: Studies have shown that there is no relationship between aging and job performance. It simply does not decline with age.

Job-related competencies are reflected in so-called "tacit knowledge," the kind of procedural knowledge helpful in solving everyday problems arising in the workplace that is not taught explicitly as part of formal training. Research has shown that tacit knowledge does not suffer any appreciable decline with age, which may explain the lack of a negative relationship between aging and job performance. In fact, tacit knowledge declines far less than the isolated mental faculties (memory, attention, and so forth) usually assessed through formal neuropsychological tests. This means that an aging professional is likely to continue to be sound on the job, despite the decline in memory and attention.

Knowledge Descriptive and Knowledge Prescriptive

Tacit knowledge is more about solving problems than about knowing facts. This brings us to a very important distinction: the difference between the descriptive and prescriptive aspects of cognition and between the descriptive and prescriptive aspects of wisdom and competence. As we pointed out earlier, knowledge can be descriptive and prescriptive. So too can be pattern recognition and the attractors that embody it in the brain.

Descriptive knowledge is the knowledge about how things are. It is sometimes called "veridical knowledge." Because things exist in the world independently of you, the observer, various statements about things can be judged as "true" or "false" regardless of your wishes and preferences. The statement "five plus five is ten" is true, and the statement "five plus five is twelve" is false. And if you wish that it were the other way around, then well, tough luck! Veridical, descriptive knowledge is the knowledge of the true nature of things.

By contrast, prescriptive knowledge is the knowledge not about how things are, but how they should be, and in particular it is the knowledge of what we must do to set them according to our wishes and our needs. Prescriptive knowledge is the knowledge of what needs to be done, the knowledge of the desired course of action. Unlike descriptive knowledge, prescriptive knowledge is not independent of you. Quite the reverse, it is knowledge about *your* needs and about the course of action that is best for *you*. Prescriptive knowledge is not the knowledge about the objective, "true" nature of things, but about the best course of action. Because the choice of such action is different for different people, I sometimes refer to it as *actor-centered knowledge*.

We humans are in command of the powerful mental machinery enabling us to acquire and store descriptive knowledge, but this machinery is secondary, ancillary, subordinated to our needs for the acquisition and storage of prescriptive knowledge. The evolutionary pressures that have shaped our brain and our body were directed at enhancing our survival and not our ability to establish the ultimate truth, even though the latter would be a nice facilitator of the former. And unless you are Diogenes living in a barrel, the primary objective for most people is to improve their lot, while finding the truth is a means to that end rather than the end in its own right.[2]

With this in mind, it comes as no surprise that prescriptive knowledge is particularly valued, and so are prescriptive wisdom and prescriptive competence. People are more likely to turn to a sage or an expert for advice on what to do than for an explanation of how things are. The prescriptive power of wisdom and the prescriptive power of competence deserve a separate discussion.

To begin with, we need to understand where in the brain knowledge is formed and stored, and also how the difference between descriptive and prescriptive expertise is reflected in the brain machinery of knowledge. And for that we need to consider two major distinctions in the architecture of the brain: the distinction between the two hemispheres and the distinction between the front and back of the cerebral cortex. Both descriptive and prescriptive knowledge are based on pattern recognition, and the patterns are embodied in attractors. Since knowledge is stored where the information was first processed (remember,

[2]Ironically, cognitive psychologists have traditionally directed most of their effort to understanding the machinery of descriptive knowledge. Only quite recently did the mechanisms of prescriptive knowledge begin to attract the kind of scientific interest that they deserve.

there is no designated, spatially separate warehouse of memories in the brain), the attractors embodying descriptive and prescriptive knowledge inhabit somewhat different neocortical territories.

Both descriptive and prescriptive knowledge are stored in the most advanced parts of the neocortex, known as the association cortex. Descriptive knowledge is stored mostly in its posterior subdivisions, in the temporal, parietal, and occipital lobes. By contrast, prescriptive knowledge is stored in the frontal lobes. Recent research has also shown that the two cerebral hemispheres play very different roles in knowledge acquisition and storage, in the formation of attractors, and in the machinery of pattern recognition.

In the next few chapters we will further explore the brain mechanisms of wisdom and competence and how these coveted traits depend on the two halves of the brain and on the frontal lobes. As we learn more about the frontal lobes, their intimate role in the acquisition and storage of prescriptive knowledge

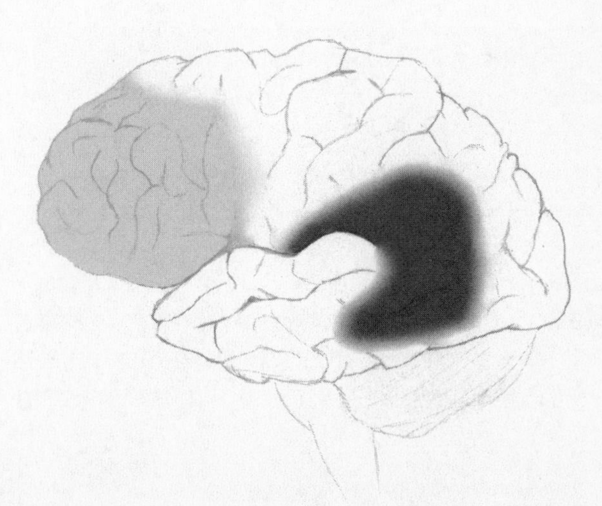

FIGURE 10. **Descriptive (darker shading) and prescriptive (lighter shading) knowledge territories.**

will become increasingly clear. And as we learn more about the differences and interactions between the two cerebral hemispheres and how they relate to new and familiar cognitive challenges, we will better understand what sets the wisdom patterns apart from other manifestations of the mind, how they come about, and what allows them to withstand the ravages of aging.

9

"UP-FRONT" DECISION-MAKING

Inside the Frontal Lobes

Today, the frontal lobes are among the most studied parts of the brain, their functions are recognized as the cornerstone of our mental world, and their changes in development and in aging are the focus of intense scientific attention. As a result, we have come to recognize the maturation of the frontal lobes as the central theme of cognitive development and their decay as the central theme of cognitive aging. But this understanding was painfully slow in coming, and one can see why: It is easier to explain what the frontal lobes do not do than to explain what exactly they do, and it took neuroscientists a long time to catch up.

I remember the first time my mother took me to the local Opera House as a little boy in my native city of Riga. I was supposed to watch the performance on the stage, but I found myself mesmerized by the little man in front of the orchestra. The little man was standing on the podium and waving his hands, and for the life of me, I could not understand just what exactly he was contributing to the performance, since he was obviously not playing any instrument. Needless to say, that little man was the conductor.

The frontal lobes, or more precisely the prefrontal cortex, are to the rest of the brain what the conductor is to the orchestra,

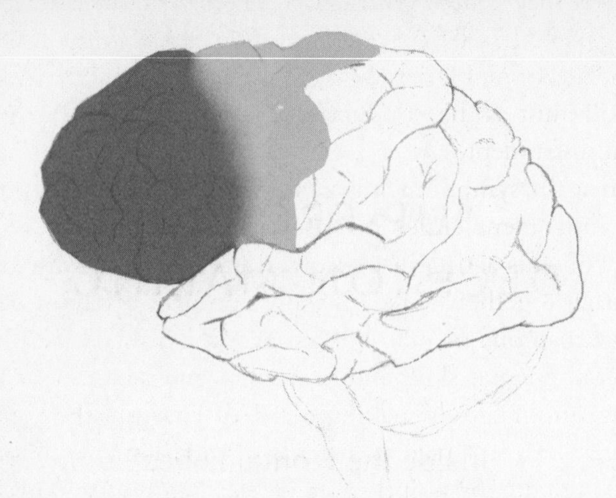

FIGURE 11. **The frontal lobe (light and dark shading) and the prefrontal cortex (dark shading).**

and for many years psychologists and neurologists were finding themselves in the position I was in as the puzzled little boy, equally unable to grasp their functions.

Paradoxically, in clinical practice the role of the frontal lobes in shaping central personality traits was recognized long ago. Frontal lobotomy, so popular in the middle of the twentieth century both in Europe and in North America, was intended to alter personality—regrettably by annihilating it in many cases altogether—by severing the connections between the frontal lobes and the rest of the brain. But the true scientific understanding of frontal-lobe functions lagged behind.

One of the main stumbling blocks was the fixation on the mechanisms of descriptive knowledge, which continued to dominate neuropsychology and cognitive neuroscience until relatively recently. As we will find out, the frontal lobes have relatively little to do with descriptive knowledge and have everything to do with prescriptive knowledge. Another stumbling block was that neuroscientists continued to study and measure specific mental skills: perception, language, movements, and so on. But the frontal lobes are not in charge of any of these specific skills, just as the conductor is not in charge of any single musical instrument.

Then again, symphonic music does not reside within any particular instrument either. It arises from the interplay of all of the instruments acting in concert. And it is the conductor who "puts it all together." Likewise, any complex behavior depends on more than one mental skill, and it is the frontal lobes that organize our mental skills into complex ensembles. The frontal lobes are in charge of making plans, of charting the paths that the organism must take in solving a wide range of problems. Like the conductor pointing his or her baton at various members of the orchestra as the music unfolds, the frontal lobes call upon specific mental skills and abilities and weave them into complex behaviors. This commanding role of the frontal lobes is often called "the executive function," by analogy with a corporate CEO, who is responsible for charting the corporate strategy but famously does not involve himself or herself with any specific, narrowly defined activity. The CEO oversees the activities of others and is in charge of identifying and marshalling the resources in pursuit of global corporate strategy. This is what the frontal lobes do in the living biological organism.

Recent evidence suggests that the internal organization of the prefrontal cortex has a hierarchical structure, indeed somewhat akin to the hierarchical structure of a large organization in society—corporate, military, or administrative—with the overall command post at the top and various divisions and subdivisions below. The very front of the prefrontal cortex is in charge of overall decision-making, and the areas of the prefrontal cortex found behind it are in charge of planning and executing the increasingly detailed subcomponents of the overall task. This is very much in keeping with the gradiental principle of functional cortical organization evident throughout the brain, as discussed earlier in the book.

The more systematic the thought processes are, the more they depend on the frontal lobes. The introduction of logical, rational method into any kind of problem-solving increases prefrontal cortical activation—as does the increase in the problem's complexity, which requires interrelating many parts and juggling many mental operations for its solution. Interestingly, inductive reasoning requires more prefrontal resources than does deductive reasoning.

The frontal lobes appear to be the engine of complex, goal-directed action and thought. This implies that the plans and blueprints of such actions are formed in the frontal lobes. So are the trains of thought underlying the rational analyses of various problems, the analytic methods that we forge in our heads as we search for the cogent approaches to these problems. They all are formed with the major participation of the frontal lobes.

We have established in previous chapters that memories for certain events are stored in the same parts of the brain that participated in the processing and analysis of those events as they unfolded. Since the plans of action and the trains of thought of rational analysis are formed in the prefrontal cortex, the memories of these plans, the memories of the past "executive" solutions for various problems, and the overall mental habits for rational analysis that one develops over time are also contained in the prefrontal cortex. Following Joaquin Fuster, we will refer to these memories as the "executive memories." These executive memories are ready for use when life repeats itself, as it invariably does, with new variations on old themes. In addition to its numerous other executive functions, the frontal lobes serve as the repository of such executive memories.

To put it in other words, prescriptive knowledge, the generic memories of the effective ways of approaching life situations and of the optimal courses of actions for whole classes of such situations, are contained and accumulated within the frontal lobes. Those in possession of such generic memories "know what to do" in thorny situations that confound others. Instead of facing every act of complex executive decision-making "from scratch," which may very well be an insurmountable task, they deal with them as with pattern recognition. In a sense, the prefrontal cortex contains the representations of future actions and of future analytic approaches applicable to situations that have yet to arise. Since both wisdom and expertise are particularly valuable for their prescriptive power, the frontal lobes are a very important part of the neural machinery of wisdom and expertise.

Much as these issues are central to the workings of the human mind, scientists are only now beginning to dare pondering them. Traditionally, certain aspects of the mind were considered the purview of the brain and thus a legitimate territory for neuroscientists to explore, and other aspects of the mind were considered the purview of the soul, the territory of poets and preachers off limits to serious, self-respecting neuroscientists. Until very recently, a decade or two ago really, cognitive neuroscience was content with dealing with such "legitimate," "bread and butter" topics like perception, movement, and memory. The more elusive and presumably "uniquely human" attributes of the mind, like motivation, judgment, empathy, insight into others, morality, and so on, were considered firmly and permanently outside the pale of mainstream scientific inquiry, and anyone purporting to bring them into the scientific discourse was regarded as a quack, charlatan, or worse. These exalted mental attributes were all lumped under the "soul department" surrendered by the scientists to the poets.

Cognitive ambiguity was also among those taboos. Received wisdom dictated that psychological experiments had to be fully deterministic. I recall a professor's admonition to us students that one "must know what the subject is doing." That meant purging the experiment of any trace of cognitive ambiguity. But most real-life situations are not deterministic—they are essentially ambiguous—and critical decision-making must occur in an opaque environment. Any experimental design failing to account for such ambiguity throws the baby of insight out with the bath water of irrelevance. Unfortunately, this was until recently the state of affairs in brain and cognitive science.

But as of late the decades-long taboos have been broken, and today the pages of serious scientific journals are replete with precisely that: exploration of the brain mechanisms of volition, drive, judgment, foresight, and decision-making in highly ambiguous situations. Even such presumably uniquely human attributes as will, intentionality, ethical behavior, morality, and empathy are being approached today with the rigorous methods of cognitive neuroscience and experimental psychology. This is reflected in the coinage of new terminology, which until recently would have been regarded oxymoronic and scandalously "snake oil–like," such as *social neuroscience* (dealing with the brain mechanisms of social interactions), and *behavioral economics* (dealing with the psychology of decision-making in the marketplace). The 2002 Nobel prize in economics went to Daniel Kahneman, a psychologist who with his late colleague Amos Tversky spent a lifetime investigating the psychological (and, as it turns out, often less than perfectly rational) mechanisms of economic decision-making in ambiguous environments.

All this is the best testimony to a new trend in neuroscience gathering steam. But the trend has gone even further. As if "behavioral economics" were not brave enough, of late we began to hear about *neuroeconomics,* concerned with the brain mechanisms of decision-making in the marketplace and using to this end the most advanced methods of functional neuroimaging. We even hear about *neuromarketing* assessing the brain responses to advertising, and about the use of functional neuroimaging to understand how political commercials work in presidential election campaigns. If one looks at these new developments closely, one cannot help but conclude that the shift of emphasis in neuroscientific inquiry is from descriptive cognition (what is true?) to prescriptive cognition (what is best for me?).

Descriptive (or veridical) and prescriptive (or action-centered) forms of cognition are closely interwoven and under normal circumstances operate in concert. Yet the difference between them is an important one, and not just for psychologists and neuroscientists. In 2002 the U.S. Supreme Court came out with what I consider a landmark opinion in the context of American jurisprudence. In ruling against capital punishment for the mentally retarded, the justices opined that an individual may possess a requisite descriptive cognition (to know rhetorically the difference between right and wrong), yet have deficient prescriptive cognition (to be unable to actually use this knowledge in guiding one's own behavior).

Traditionally, neuropsychologists and cognitive neuroscientists focused almost exclusively on the brain mechanisms of descriptive, veridical cognition. It is only very recently that they have turned toward the understanding of prescriptive, actor-centered cognition. Because it can be cogently argued that evolutionary pressures shaping the design and capacity of our brains are primarily about finding "the course of action that's best for me" and only secondarily and derivatively about "finding the truth" (the latter being clearly in the service of the former), it is ironic that it took neuroscientists so long to turn their attention to the brain mechanisms of prescriptive cognition, but it finally happened. Better late than never!

In this new kind of inquiry the focus is on the frontal lobes of the brain, since they contain the neural machinery of prescriptive knowledge. The frontal lobes, the most recent to develop in evolution and situated in the driver's seat in front of the rest of the brain, have also been the one part of the brain to guard their secrets most jealously. But as neuroscientists finally began to tackle prescriptive, actor- and action-centered cognition, it became clear that such cognition essentially depends on, and is driven by, the frontal lobes. The role of the frontal lobes in prescriptive knowledge has been elucidated in our research into the functions of the frontal lobes using functional MRI conducted in the laboratory of Kai Vogeley.

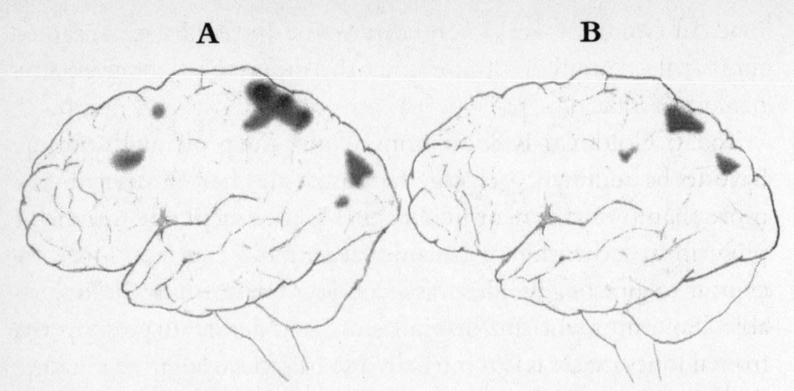

FIGURE 12. **Brain activation in actor-centered (A) and veridical (B) tasks on fMRI.** *(A) Choosing geometric forms on the basis of preference produces combined prefrontal and parietal activation. (B) Choosing geometric forms on the basis of perceptual mismatch produces only parietal activation. Adapted with permission from Vogeley et al. (2003).*

Cinderella and the Brain

If there is such a thing as the change of fortunes for a brain region, then we have certainly witnessed it for the frontal lobes, in their transition from the Cinderella of neuroscience to one of its hottest topics. Even as relatively recently as the middle of the twentieth century, the frontal lobes were thought by many scientists to exist for merely ornamental purposes, or at most to serve the function of supporting the cranium lest it collapse. (This was the case despite the prescient admonitions of such scientific visionaries as John Hughlings Jackson and Aleksandr Luria, who anticipated the exceptional importance of the frontal lobes in human cognition.)

I recall Patricia Goldman-Rakic, one of the most important students of the frontal lobes, showing the outline of the brain in a lecture given at Columbia University many years ago. Every lobe contained a certain number of homunculi, tiny stick men, reflecting the amount of scientific interest directed toward that lobe. In Goldman-Rakic's rendition, the frontal lobe contained the smallest number of homunculi. It was the Cinderella, the neglected lobe.

Today, Goldman-Rakic's homunculus map would certainly have to be redrawn, a change to which she herself contributed more than most. As neuroscientists began to tackle the secrets of prescriptive cognition, it became clear that the frontal lobes are central to practically all its aspects, omnipresent and indispensable. So important and overarching is the role played by the frontal lobes that it is often referred to as "meta-cognitive" rather than merely cognitive. To be precise, not the whole frontal lobe has been implicated in these highest-level aspects of the mind, but rather its one particular portion, the prefrontal cortex.

The ascent of the prefrontal cortex in evolution has been relatively recent, reaching an appreciable level of development

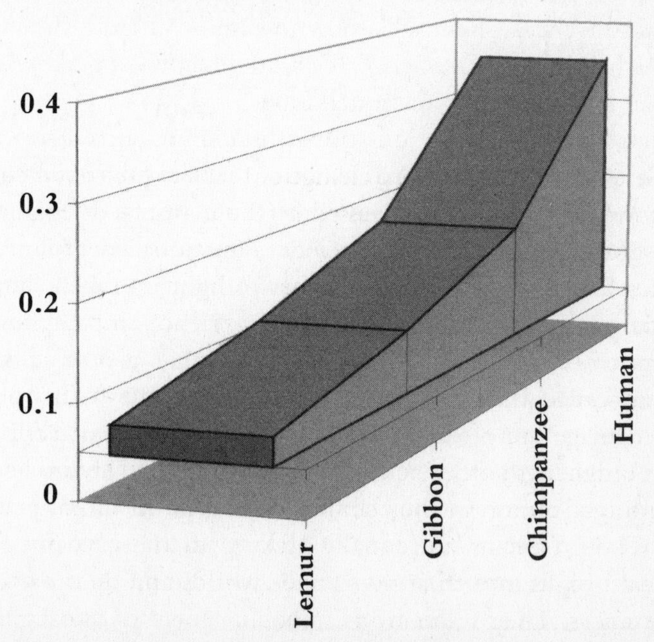

FIGURE 13. **Evolution of the frontal cortex.** *Expressed as the ratio of frontal cortex to all cortices. Based on Brodmann. (1909).*

only in the mammalians and exploding exponentially only in the higher primates. This is consonant with the exceptional role that the prefrontal cortex plays in the mental attributes that are often regarded, rightly or wrongly, as uniquely human and set us apart from our fellow mammalians.

As the pendulum of neuroscientific interests swings and an ever-increasing amount of attention is being lavished on the frontal lobes, it has become fashionable to implicate the prefrontal cortex in virtually everything representing the highest strata of human spiritual attainment, and to resort to extravagant language in describing its function. An authority no less august than Aleksandr Luria himself referred to the prefrontal cortex as "the organ of civilization," and it has also been said that the whole human part of evolution was dominated by the evolution of the frontal lobes. I, too, am guilty of contributing to this mythology by putting on the jacket of my earlier book a tongue-in-cheek, slightly blasphemous spin-off from the image of Michelangelo's *Creation of Adam,* in which God makes Adam human by lighting up his frontal lobes.

Such excessive exaltations notwithstanding, current research has shown beyond a reasonable doubt that the prefrontal cortex is important, perhaps even central, to those aspects of cognition, that glue individuals into society. Functional neuroimaging studies have shown the frontal lobes to light up when subjects ponder moral or social dilemmas, or experience empathy toward others, or when they are asked to "read other people's minds" (the so-called theory of mind studies). Patients with frontal-lobe damage are often astoundingly lacking in their ability to form insight into other people's inner worlds and are no less astoundingly obtuse in their ability to engage in moral reasoning. These patients are equally lacking in the capacity for a critical insight into their own inner worlds and their own circumstances. They suffer from a peculiar form of anosognosia, different in character from the anosognosia caused by right-hemispheric damage, but no less devastating. And research has shown that criminals, particularly violent criminals, often have abnormally small or abnormally physiologically inactive prefrontal cortices. Reduced volume of prefrontal gray matter has been found in people with antisocial personality disorder, and an insufficient level of frontal activation, usually referred to as *hypofrontality,* has been found in people prone to impulsive aggression.

Does this mean that people are born with certain "moral knowledge," or "social knowledge," which resides in the frontal lobes? Has our naive search for a "module for every thing" finally culminated in the discovery of a "morality module" containing a "moral instinct?"

Indeed, it has become fashionable to talk about the frontal lobes as "the seat of morality." Is there such a thing? I doubt it. Knowing what we know about the history of human civilization, I am rather skeptical about any notion of inborn morality. Taking a position, which is neither overly romantic nor overly nihilistic, I tend to think of the brain as a morally agnostic device, at least in a literal sense. To me, any notion of "the morality instinct" sounds every bit as fantastic as the notion of "the language instinct," or even more so. I strongly believe that the ethical norms regulating our social behavior are by and large cultural rather than "hardwired" constructs.

Does this mean that I am inclined to deny the prefrontal cortex any role in moral development? Not at all! I do believe that the prefrontal cortex is critical for the formation of ethical concepts, but only indirectly. We know that the prefrontal cortex is responsible for the "sequential organization" of behavior, for organizing behavior in time, and for arranging the various mental operations that go into any complex act of cognition into temporally ordered and coherent sequences. This most probably means that the prefrontal cortex contains the brain mechanisms of establishing the relationship between "before" and "after." Then, by virtue of its ability to establish *temporal relations*, the prefrontal cortex has become critical for the next level of abstraction, for establishing the more complex *causal relations*, the relations between causes and consequences.

A well-developed prefrontal cortex is probably necessary for the establishment of a whole class of relations of the "if then" ("if A then B") type. The ability to grasp this relationship as a general rule is present in humans but is absent in nonhuman primates. This ability is the cornerstone of a number of complex cognitive skills, which, rightly or wrongly, we associate with humans. Language is one of them, since the "if then" structures are the basis of complex grammars that serve as the foundation of complex language. This highlights the often-overlooked role of the prefrontal cortex in the emergence of language in evolution, language development in children, and the everyday use of language.

But the capacity for grasping "if then" relations is likely to be at the heart of moral development as well. While not in itself sufficient, the ability to interrelate causes and consequences is necessary and serves as the prerequisite for any moral reasoning and for the grasp of ethical concepts. To turn my earlier statement around, even though the frontal lobes contribute to moral reasoning only indirectly and are intrinsically morally agnostic, they do provide the critical building block, the neurobiological cornerstone on which the development of these concepts rests.

Foremost among the other foundations of moral reasoning is the ability to imagine the consequences of the alternative courses of actions—"what would have happened had I done X instead of Y"—and the capacity for regret when you reach the conclusion that having found yourself at a decision-making crossroads you took a wrong turn. The capacity for such "counterfactual reasoning" is important not just in the moral sphere. It is equally important for sound decision-making in every arena—economic, political, or personal. Deprived of the benefits of counterfactual reasoning, any "learning from experience" would be sorely handicapped and reduced to trial and error. Today we know that the capacity for counterfactual reasoning and for the experience of regret depends on the frontal lobes. As a group of French scientists led by Nathalie Camille have shown, patients with damage to a certain aspect of the frontal lobes—the orbitofrontal cortex—lose these abilities to a considerable degree.

Empathy, insight into the minds of others, and the capacity for moral reasoning are among the most important ingredients of wisdom by any definition, on a par with the capacity for effective problem-solving. According to many definitions, wisdom implies the ability to integrate pragmatic "actor-centered" and ethical "empathy-driven" considerations, and this agrees with my own intuitive sense of the essence of wisdom. The unique role of the prefrontal cortex lies in its providing the neural machinery for bringing these two factors together in a single, well-integrated decision-making process.

Today, there are many reasons to believe that the development of the prefrontal cortex has played a central role in the emergence of many (and possibly of most) traits that we recognize as defining our humanity as a species. Does this mean that these traits are *uniquely* human? Whatever the true degree of evolutionary discontinuity between the human brain and those of other species may be, we often tend to exaggerate it in our romantic (self-serving) view of ourselves. Some restraint is necessary in order to not anthropomorphize certain traits, which may very well culminate in humans but are almost certainly not unique to us in a qualitative sense, and are not strictly speaking dichotomous, characterized either by their absolute presence (in humans) or by their absolute absence (in every other species).

Take "empathy" and the ability to "read another person's mind." These admirable traits, known in neuroscientific parlance by the unwieldy term "the theory of mind" capacity, are undoubtedly necessary to glue any human group together. Functional neuroimaging research has shown that these lofty mental traits depend on the frontal lobes. We proudly appropriate these mental gifts to ourselves, humans, the *Homo sapiens*

sapiens, and are loath to grant them to other species. In the face of compelling evidence, we begrudgingly concede that certain primates others than ourselves, namely the great apes, possess the rudiments of this capacity. A famous photograph comes to mind of a baby chimpanzee running to comfort its human handler feigning distress.

But how about my bullmastiff Brit? In his puppyhood he, as every puppy, liked to sneak into my bedroom closet, grab a sock, dash out, and settle on the living room couch chewing the sock with abandon. Like every puppy owner, I gave chase and took the sock away from him. But after a while, Brit's behavior had changed. He still raided the bedroom in search of my socks, but now instead of eloping with them to his favorite couch, he began to bring them directly to me, wherever in the apartment I happened to be at the time.

I never taught Brit this behavior. And yet with his canine mind, he must have surmised that if I always took the socks away from him, I must have needed them badly, and with his good canine heart he obliged by bringing them directly to me. By any common-sense definition, Brit exhibited a certain capacity for "the theory of mind!"

What's more—something I discovered by sheer accident—when I feign distress by burying my face in my hands and engaging in make-believe crying, Brit becomes alarmed, interrupts whatever he is doing at the time, runs toward me, and licks my face. He does not do this when I feign other emotions, pretending to laugh happily or yell angrily. He seems to be quite selective in differentiating between my emotional states, although we are not even of the same species (but admittedly great friends). Canines are nowhere near the pinnacle of frontal-lobe development, yet they undoubtedly have some frontal lobes, presumably enough for Brit to manifest rudimentary (perhaps even not so rudimentary) ability to "read my mental states" and express empathy.

And when I am absorbed in an animated conversation with a

fellow human, reading a book, or working on my computer, after a while Brit begins to paw me gently but persistently. He appears to do so not because he needs anything in particular, but because he wants my attention to be directed at *him*, and I am tempted to see this as evidence of rudimentary self-awareness, another cardinal attribute of developed mental life presumably dependent on the frontal lobes.

All my affection for Brit notwithstanding, I don't believe that he is a "superdog" in possession of unique mental attributes absent in other canines and other mammalians. I am convinced that if I had a different creature, a pack animal of a comparable evolutionary standing for a pet whom I raised from early puppyhood as I did Brit, I would have made similar observations. So perhaps it is wrong to think that any of these exalted attributes of developed cognition—empathy or the theory of mind—appeared precipitously, like a deus ex machina, at the very final stages of evolution. More likely, they developed gradually and incrementally through much of mammalian evolution, and this process began as soon as the frontal lobes made their first appearance on the evolutionary scene.

As befits the orchestra conductor or a corporate CEO, the frontal lobes are exquisitely well connected. Or to amplify the importance of connections, think of a puppeteer (the frontal lobes), whose ability to control the puppets (other parts of the brain) depends on the strength of the strings, without which the puppeteer will be completely powerless. As we already know, these strings, the pathways connecting the frontal lobes with the rest of the brain, are extremely slow to mature: They reach their fully operational state only sometime between the ages of eighteen and thirty. This was established by studying the time course of the appearance of myelin, the white fatty tissue that insulates the pathways and by so doing enhances the speed and reliability of signal transmission along the pathways.

Another way of ascertaining frontal-lobe maturation and its evolutionary course is by studying the development of so-called "spindle cells." Spindle cells relay information across very far-flung regions of the brain and are found predominantly in the orbitofrontal cortex. These cells begin to appear during the first months of life, but their number increases drastically during the second and third years of life. The spindle cells are particularly intriguing, since they are very prolific in humans, modestly represented in great African apes, and are completely absent in other species. This makes it tempting to link the spindle cells to consciousness, volition, and other highly advanced attributes of the mind.

Let's now step back and ponder again the social meaning of the eighteen to thirty time frame. As we concluded earlier, the age of eighteen is a very interesting age, recognized by most modern Western societies as the age of social maturity, of passage from adolescence to adulthood. The age of thirty (again give or take a few years) is also a very interesting age. This is the age when people are allowed to hold high elective offices in various Western societies. So it appears that even without the benefit of explicit neuroscientific advice, numerous modern societies have "discovered" that the period between eighteen and thirty years is when the critical aspects of social maturity finally fall into place. As I already argued, the agreement between the chronologies of social and biological maturation of the frontal lobes is hardly coincidental. Today, many scientists (I am one of them) believe that the mature and fully functional frontal lobes are the prerequisite for social maturity.

Of course, this does not mean that executive functions appear on the scene at a certain age precipitously and abruptly and that an instantaneous transition from their complete absence to their full-fledged presence occurs at a certain age. Like most biological and cognitive attributes, they develop gradually, so at any stage of cognitive development the correct question to ask is not "yes or no?" but "how much?"

The same consideration applies to individual differences between people. Like every other attribute of our mental life, the executive functions, the conductor's prowess that depends on the frontal lobes, differ from person to person. The neuropsychology of individual differences is still in its infancy, but we widely accept the notion that perfectly normal, upright citizens vary in their musical talent, literary talent, athletic talent, and every other talent. Even though in everyday life we often judge some people as "having it" and others as not, it is usually not a matter of "yes or no," but rather of "how much."

Like every other aspect of our material self—our height, our weight, the color of our eyes and hair, our lung capacity, and typical blood pressure—our brains are also subject to individual differences. Exactly "how much" of various talents one has depends to a huge extent on the individual differences that characterize our brains.

This general premise applies to the executive functions as well. The recently popular term *emotional intelligence* captures some of that which the frontal lobes control, but not all. To the extent that we find the concept of "executive functions" useful (and we do), the concept of "executive intelligence" is probably also in order. The various contributions of the frontal lobes, such as planning, foresight, capacity for impulse control, empathy, and "theory of mind," tend to represent a cohesive package. They travel together: in any given neurologically intact individual they tend to be all well developed, all modestly developed, or all poorly developed. The term *executive intelligence* will encompass all these functions of the frontal lobes and reflect their cohesion. And it would exhibit the same degree of individual variation as any other aspect of our mental life.

Any complex real-life situation requires deployment of the executive functions controlled by the frontal lobes, probably not all of them in the context of any single task, but a significant subset thereof. In order to better understand how the frontal lobes guide behavior in real life, imagine an advanced middle-age man trying to write a book. He is not a professional writer and the language in which he labors is not his native tongue. This makes the process somewhat more deliberative than it might have otherwise been, and to make up for his obvious handicaps, our author relies heavily on his frontal lobes.

Writing styles differ. I have heard from several authors that they don't know what will come out from under their pen until they actually put the pen to paper (or fingers to the keyboard). In this approach thinking and writing are intertwined in a single fluid process. But our imaginary author goes about his business in a very different manner. He plans before he acts. Every day he spends a fair amount of time in what might appear to be an exercise in forced idleness, circling Central Park with his very big but very friendly dog. But he is not idle. He thinks through the outline of his book and its various chapters long before he actually writes a single word. He creates an overall plan first and acts according to it later, and in so doing he deploys his frontal lobes. Since the ability to create a plan of action ahead of the action itself is linked to the prefrontal cortex, it must be abuzz in our author as he ambles with his dog from Strawberry Fields to Bethesda Fountain. This writing style results in a very peculiar process, which is more akin to sculpting than to writing. Our author first creates a general outline in his head, then writes a skeleton draft of the chapters, and only then embellishes every chapter bringing it to a relatively finished form. The process is not linear, from chapter one to chapter two to chapter three and so on. Instead, the process is *hierarchical*: from a very general outline to a collection of skeleton chapters to finished chapters. The process is almost more architectural than literary. The ability to engage in parallel activities through a top-down unfolding of a plan is also controlled by the frontal lobes.

By its nature, the book in progress is a mix of several subjects, including biology, psychology, and history. This means that for each section of the book the author must access a particular part of the knowledge bank that he has accumulated in his fifty-something years of life. While he does this relatively automatically and effortlessly, the act of mental selection is guided by his frontal lobes, which are kept quite busy by this process.

The author definitely wants the future reader to finish the book, which means among other things that it should not be too long. This further strains the selection process. Our author cannot include all his relevant knowledge in the book and must somehow prioritize. In so doing, the author deploys an internal editor of sorts, a gatekeeper whose function it is to allow certain items of information into the book and reject others as less important. More strain on the frontal lobes, since they are that internal editor.

As our author moves from chapter to chapter, the topics shift from biology to psychology to history, then back to biology and so forth. That the author is able to switch from one topic to another in a relatively seamless way is further tribute to his frontal lobes, since the prefrontal cortex is in charge of mental flexibility.

Like most writers, our author aspires to saying something new, something original, something not said or written before. He is trying to create new content. But very few things are totally new. In most cases, new content is somehow related to old knowledge. How is such new content created? It is created by configuring in novel ways bits and pieces of old knowledge. The elements are old but their configuration is new, without precedent in the past and thus not corresponding exactly to any of the previously formed mental representations already contained in the author's head. Assembling a new mental configuration out of the elements of old mental representations is very different from merely accessing old mental representations, as would be the case, for instance, in listing established facts in a review paper or an encyclopedia. Again, the prefrontal cortex steps in, since it plays the unique role of "working with mental representations," rearranging and reconfiguring them according to new needs.

Every writer wants the reader to enjoy his book, to find it

interesting and enlightening, and our imaginary author is no exception. To accomplish this, he must enter the reader's mind, to put himself into the reader's mental shoes, to form in his own brain a mental representation of the reader's inner world. When an author deletes a paragraph because it is "boring" or spices up the narrative with an anecdote because it is "funny," he makes these assertions from the reader's point of view. He attempts to read his putative reader's mind. As we already know, this ability rests firmly with the frontal lobes.

Finally, the manuscript is ready—sort of. Fortunately for our author, he has an excellent editor who goes over the manuscript with a constructive yet critical eye. This external editorial input is also guided by the frontal lobes—in this case the editor's frontal lobes.

The Frontal Lobes and the Aging Mind

It may be a good thing that the editor is more than twenty years younger than our imaginary author, since unfortunately, as we already know, the frontal lobes are more susceptible to the effects of aging than most other parts of the brain. But the frontal lobes' susceptibility to decay does not automatically mean that prescriptive wisdom, or prescriptive expertise, disappears with age. Nor does it mean that the frontal lobes decay uniformly in everybody. Joaquin Fuster proposed that the prefrontal cortex contains the generic representations of schemes of actions effective across whole ranges of situations and problems. He refers to them as *semantic executive memory* and *memory for concepts of actions.* Since these memories may differ in their generality, they are hierarchically organized. Like other generic memories, executive generic memories are relatively invulnerable to the effects of brain damage. This enables an aging corporate CEO or an aging political leader to remain an effective executive, despite some cognitive decline in the ability to engage in de novo mental computations. (Think of Winston Churchill with his attention drifting but his strategic grasp unassailable.)

Those who are endowed with "executive intelligence" have a considerable leg up in finding the optimal course of action in genuinely novel situations. Consequently, as they move through life, they are likely to accumulate a large "neural library" of "generic executive memories," the memories of past successful solutions to thorny problems, in the form of attractors residing, completely or in part, within the frontal lobes. Their neural library will be more extensive than that of most people. As a result, they are particularly likely to find effective "executive" solutions for seemingly new thorny situations by way of honing in on their similarities with some of the old, previously solved problems through rapid executive pattern recognition. While not identical, "emotional intelligence" and "executive intelligence" are closely interrelated. To the extent that emotional intelligence also has a distinct seat in the brain, the frontal lobes are that seat. And the executive memories stored within the frontal lobes are informed by emotional intelligence.

Like every other aspect of aging, the rate of frontal-lobe aging is subject to individual differences. A greater functional longevity of the frontal lobes is probably an important key to a sound mind in advanced age. Those who preserve a good working condition of their frontal lobes are the ones with the best chance of remaining clear-minded well into old age. Indeed, it has been shown that high-functioning elderly individuals have more physiologically active frontal lobes.

Even when it happens, the decay of the frontal lobes in aging is likely to affect mostly the ability to find "executive" solutions for genuinely novel situations. But what is novel and what is familiar also varies from person to person. Since most new situations resonate to some degree with previous experiences, people who have accumulated a vast neural library of well-entrenched generic executive memories are likely to remain effective problem-solvers even despite this decay, at least for some time.

In the forthcoming chapters we will discuss how mental activity and mental exertion actually strengthen the underlying neural tissue. This is true also for the frontal lobes. So the people with a lifelong history of complex executive decision-making are more likely to preserve the neural integrity of their frontal lobes well into old age than passive "follower" types with relatively modest exertion of their executive function earlier in life.

Moving executive talents and executive intelligence from the domain of the Platonic soul to the domain of the biological brain is fraught with important implications. In an interview with the *Harvard Business Review* some time ago, I was asked if the executive talent can be developed and how. This question, obviously of particular interest to senior corporate executives, should be of interest to the general public as well. Regardless of what we do for a living, we are all faced, to a greater or lesser degree, with "executive" decisions in the context of managing our individual lives.

My response was reserved. Perhaps the executive talent can be developed, and it is important to figure out how this can be done. But it is as important to be able to recognize it when it is naturally present in some people. And it is equally important to recognize its natural absence in other people. Like every other biologically grounded attribute of the mind, the gift of executive talent is not awarded to everyone in equal measure. Instead of tacitly embracing the assumption that executive skills can be developed with equal ease in everyone, corporate leaders should do what athletic coaches, choreographers, and music teachers have been doing for as long as they have been teaching students their secrets: look for natural talent and focus their mentorial energies on those naturally endowed, instead of squandering them on every comer. They know that astute selection is key to success more than anything else.

This, of course, begs the next question: How does one look

for the executive talent? Again, it is easier to say how *not* to look for it. Not by giving the IQ test, for instance. It has been shown that the most successful top corporate executives don't usually have exceptional IQs. Their IQs are respectable, within what is called "high average to superior" range. But they are not off the scale, not even close. In the same vein, patients with severe frontal-lobe damage (due to stroke, brain trauma, or other neurological conditions) frequently have normal IQs, despite the fact that their capacity for meaningful behavior is completely devastated.

The reader of this chapter, who has navigated his or her way to this page, is probably already sufficiently impressed with the complexity of executive functions. Given its multifaceted nature, it may be neither possible nor practical to measure it with a single yardstick. Multiple measures may be required, examining separately such attributes as planning abilities, mental focus, mental flexibility, empathy, the ability to deal with novelty, and the ability to place yourself in another person's "mental shoes."

The ability to place oneself in another person's mental shoes is particularly interesting. Although, as we have established earlier, even my dog Brit has a modest amount of this capacity, an individual interacting with, and directing the activities of, scores of other people must be endowed with a particularly generous dosage of this gift. The ability to penetrate the other people's minds is equally essential in altruistic, cooperative, and adversarial situations. You must be equally able to have some insight into another person's mind in order to be a good friend or an effective competitor. People who have lived long, successful lives usually have had their share of both kinds of interactions.

Either way, the ability to penetrate other people's minds begins with an interest in other people's minds. The importance of this statement is hard to overestimate. I truly believe that an interest in other minds is among the foremost prerequisites of executive intelligence.

I do not purport to know any particularly pithy ways of measuring this mental attribute, the curiosity in other people's minds. But I do believe that it lends itself very well to naturalistic observation. In the company of other people, does one always engage in a self-indulgent soliloquy or does one ask questions, at least occasionally? In my scheme of things, the latter type of behavior holds out the promise of executive intelligence. But the former type of behavior brands one as a hopeless executive dummy, no matter how pompous or full of oneself.

In my own experience, the people who impress me as being particularly insightful and shrewd tend to extract the maximum amount of information from you, rather than dazzle you with the displays of their own knowledge or acumen. But we have all been in situations when such self-indulgent traits have been displayed to a comical degree. I have witnessed ignorant people expounding on their worldview in the presence of seasoned diplomats, their ideas about music in the presence of accomplished musicians. And I have had my share of people pontificating before me, with an aura of authority, about the fate of Russia, based on their five-day-long packaged tour of the land. What a waste of time for all the parties involved! Though how diagnostic!

10

NOVELTY, ROUTINES, AND THE TWO SIDES OF THE BRAIN

The Riddle of Duality

The enigma of the frontal lobes is an example of how stubborn the brain can be in guarding its secrets. But no riddle of brain organization has attracted as much attention over the years—in both scientific literature and lay press—as the riddle of brain duality. Why does the brain consist of two halves, the left and the right hemispheres, and how are they different? This question is central to our narrative and to the understanding of the mental traits that this book attempts to elucidate.

Genius and wisdom, talent and competence, are the twin gifts, equally revered yet very different. As we already know, these gifts are not joined at the hip and one may be present without the other. We also know that their peak expressions correspond to different ages: genius and talent tend to reveal themselves in youth, and wisdom and competence at the later stages of life.

What is the brain machinery of these two kinds of gifts, so closely linked yet so separate? How are they contrasted, and how are they connected in the brain? We are finally ready to tackle this question.

As discussed, competence and its supreme form wisdom depend on the availability of patterns containing both descriptive and prescriptive information. These patterns enable us to recognize seemingly unique thorny problems as variants of previously encountered problems, of problems already solved.

But what happens when you face a situation that really does not fit, even remotely, any of the patterns stored in your brain? Based on an earlier discussion, we already know that pattern formation is an intricate and lengthy process that cannot be truly understood in binary "yes/no" terms. This means that a pattern may be partly formed and partly ready for use. We will get to these subtleties later, but for now, for clarity's sake, let us consider a simplified range of possibilities. Suppose that when a person encounters a problem, a relevant pattern is either found in that person's cognitive repertoire or is not found. Or to use Stephen Grossberg's terminology, *adaptive resonance* with one of the previously formed attractors occurs or does not occur. Now we have two classes of situations to consider: familiar and novel. How does the brain deal with these two kinds of challenges?

Enter cerebral hemispheres and the mystery of duality. Duality is one of the most fundamental and universal properties of the brain. It permeates all its levels, from the brain stem to the neocortex. For every structure, nucleus, and pathway, there is a twin. In the past we used to think that this duality was characterized by perfect symmetry. Today, we know that the brain's symmetry is only rough and partial. True enough, the brain is more symmetric than asymmetric; the two halves of the brain are better understood as two variations on the same fundamental theme than as two entirely different themes. True also, that the two halves of the brain don't operate in isolation from each other. They are connected with rich pathways at every level, both cortical and subcortical. At the cortical level, the pathways connecting the two hemispheres are organized in a large structure called the *corpus callosum* and *anterior* and *posterior comissures*. These and other pathways ensure an ongoing cross-talk between the hemispheres, or rather a myriad of parallel cross-talks.

Ultimately, then, the brain operates as a well-integrated whole and not as two disjointed parts. But this unity turns out to be a unity of contrasts. As we will see, the subtle structural and biochemical differences separating the two halves of the brain translate into profound functional differences between them.

Among the very few parts of the brain that escape the duality imperative are the endocrine glands, pineal and pituitary, two small collections of nuclei buried deep in the middle of the brain. It was the pineal gland's unique status of singularity, as opposed to duality, that compelled the great seventeenth-century philosopher René Descartes to declare it the place where the body and soul meet, thus hoping to resolve the dilemma of his own creation, the body-mind dualism. Today, we know that the pineal gland plays a much more humble, yet by no means unimportant, role producing melatonin and helping regulate the sleep-wake cycle. Another structure that escapes the duality imperative, the pituitary gland, plays a role in the secretion and release of various hormones.

Few topics have captivated more attention and have inspired more outlandish speculations, than the duality of the human brain. Why do we need two halves of the brain? Why is two better than one? Numerous theories and hypotheses have been aired over the years to address these questions, but invariably contrary evidence emerged challenging them, and often annihilating them altogether.

Language and the Brain: The Roots of Misconception

The quest for understanding the functions of the two sides of the brain has been traditionally dominated by several tacit assumptions. The first assumption was that the differences are limited to the cortex, the so-called cerebral hemispheres. The second assumption was that these differences concern only brain function and that the structure and biochemistry of the two sides of the brain are the same. The third assumption was that the differences between the two sides of the brain exist only in humans and that in all other species the two sides of the brain are structurally, biochemically, and functionally symmetric.

As it turned out, all of these three assumptions obscured the picture rather than clarified it, and all were ultimately proven wrong. This, in turn, has forced the revision of one of the most entrenched tenets of neuropsychology and cognitive neuroscience: that the distinction between language and nonverbal functions captures the essence of the difference between the two sides of the brain.

To get to the root of this misconception, we need to examine certain basic facts about language and the two sides of the brain. It has been known for years that the left cerebral hemisphere plays a greater role in language than the right hemisphere; hence the term *language-dominant hemisphere*. The evidence to support this position was aplenty. In adult patients, stroke, traumatic brain injury, or any other kind of brain damage disrupts language, producing a condition known as "aphasia" when it affects the left hemisphere but not when the right hemisphere is disturbed. (As we will soon find out, in children the picture is far less clear-cut, a circumstance with far-reaching implications, whose importance eluded the theorists of hemispheric specialization for years.)

Electric stimulation of the left temporal lobe during neurosurgery produces verbal hallucination–like experiences: The patient literally hears voices saying intelligible words or even phrases. Auditory hallucinations so common in schizophrenia also usually have the appearance of well-formed utterances rather than unintelligible sounds. This probably reflects the fact that the left hemisphere is more affected in schizophrenia than the right hemisphere. A seizure focus located in the left temporal lobe produces similar hallucinatory experiences of hearing voices (which is why temporal lobe epilepsy is sometimes misdiagnosed as schizophrenia). Dyslexia, a disorder of language development in children, is more common among left-handers than among right-handers, possibly reflecting early damage to the left hemisphere and the consequent switching of handedness (a phenomenon often referred to with the unkind term *pathological* left-handedness, as distinct from *natural*, hereditary left-handedness). If aphasia (impediment of language) is caused by left-hemispheric dysfunction, then the peculiar condition of *hyperphasia*, a parrot-like mastery of long verbal scripts commonly observed in Williams syndrome, is associated with a larger-than-normal size of the left hemisphere. All this evidence points to the left hemisphere as the "seat" of language. Again, the evidence has been mostly limited to adult patients, thus distorting the overall picture of hemispheric specialization and masking some of its very important aspects.

By contrast, damage to the right hemisphere was thought to impair mental processes that do not depend on language, such as the disturbance of facial recognition (a condition known as "prosopagnosia") and impaired appreciation of music ("amusia").

These and other, similar findings have shaped the prevailing assumptions about the fundamental nature of the functional differences between the two sides of the brain. Because of the paramount role of language in human society, the term *language-dominant hemisphere* was condensed into simply *dominant hemisphere*, thus implying a somehow greater importance of the left half of the brain. By contrast, the right hemisphere was often termed *subdominant hemisphere*, implying its lesser status and presumably greater dispensability. Even today, neurosurgeons tread very cautiously when they operate on the left hemisphere, but are often much more cavalier when it comes to the right hemisphere.

Note that the important contrast deemed to capture the fundamental differences between the hemispheres is not between acoustic and visual information, but between the processes based on language (spoken and written alike) and the processes that do not involve language (auditory and visual alike).[1]

But even this (as we are about to see) simplistic understanding was further reduced to a sound bite, an oversimplification of oversimplification, so to speak. Gradually, the belief developed that the left hemisphere is fundamentally the language hemisphere and the right hemisphere is the visuospatial hemisphere. This belief, in its literal form, is still shared by many scientists studying the brain and probably by most clinicians, psychologists, and physicians treating brain disorders, since it usually takes years for state-of-the-art knowledge to trickle into the clinical trenches. But it is patently wrong. New scientific evidence challenges this assertion and forces us to adopt a totally new way of understanding the duality of the brain.

Without going into too much technical detail, let me explain why. In our pursuit of hemispheric differences, as in many other pursuits, clarity of thought and Logic 101 often turn out to be better guides than arcane technical knowledge. Logic 101 dictates that any distinction, based on contrasting language and nonverbal mental processes, is meaningful only for creatures endowed with the power of language. We humans are the only species endowed with this power, at least in the narrow definition of language. Therefore, we are the only species for which the distinction between language and nonverbal functions has any meaning. As it turns out, this inevitable conclusion creates a huge theoretical and empirical problem, nothing short of a death knell for the language–visuospatial theory of hemispheric specialization.

---

[1]The failure to understand that spoken and written language have fundamentally the same roots and are mediated to a large extent by the same brain structures is still rampant. This leads to all kinds of misconceptions in the world of education and learning disabilities. Particularly common, for instance, is the ignorance of the fact that most dyslexias (reading disorders) are secondary to dysphasias (spoken language disorders). This, in turn, leads to misdirected diagnosis and remediation.

Indeed, the prevailing assumption for many years was that functional differences between the two parts of the brain existed only in humans. The assumption was logical and made sense, at least on the surface. But another assumption, equally entrenched for many years, did not make sense even on the surface, and it was regarded as unsatisfactory by a number of scientists. This implausible assumption was that in humans the two brain hemispheres were the structural and biochemical mirror images of each other. This notion has bothered scientists for an obvious reason: Functional differences between the two hemispheres must have some material basis. The prevailing assumption of their perfect structural symmetry was inconceivable and ran afoul of common sense by implying that two identical structures would give rise to two vastly different sets of functions.

Prompted by a feeling of disquiet and bolstered by the arrival of powerful neuroimaging technologies, a number of top neuroscientists embarked on the search for the structural differences between the two hemispheres capable of accounting for the functional differences. Since the impetus behind the search at the time was to explain the link between language and the left hemisphere, the focus was on the "language areas" of the brain. Much of the early work focused on high-precision measurements of the cortical language areas and was conducted by Norman Geschwind (arguably the father of North American behavioral neurology) and his associates.

It didn't take long for the structural differences between the hemispheres to start revealing themselves. Two brain regions are particularly important in language: the *planum temporale*, in charge of speech sound discrimination, and the *frontal operculum*, critical for speech articulation. Both turned out to be larger in the left hemisphere than in the right hemisphere in right-handed individuals. What could be a better explanation for the left hemisphere's supremacy for language?

But very soon it was discovered that these structures are also larger on the left side than on the right side in great apes, who have no "language" (despite the famous Koko, trained by scientists to use simple sign language in the 1980s). What's more, paleoanthropology tells us, through the studies of endocasts (imprints on the inner surface of the skull), that already the Australopithecus had an asymmetric brain. As the search continued, numerous other differences between the two hemispheres were found, involving both brain morphology and brain biochemistry. And none of them proved to be unique to humans. Instead of setting us apart, all the differences between the two sides of the brain unite us with other species in one family. We share most of them with other primates and some of them even with such humble nonprimate mammalian species as rats and mice.

The differences between the two sides of the brain turned up on every scale of observation in both the human and nonhuman brain. This included every level, from the aerial view of the brain ("gross neuroanatomy"), all the way down to the molecular level. At the level of gross neuroanatomy these differences include a greater frontward protrusion of the right hemisphere and greater backward protrusion of the left hemisphere (the so-called "Yakovlevian torque"), and differences in the size of the planum temporale and the frontal operculum (both larger in the left hemisphere). At a finer level of brain wiring, disparities have been found between the cortical thicknesses of the two hemispheres (thicker on the right than on the left, at least in males). At the finest level of microwiring ("cytoarchitectonics," in scientific parlance), differences have been found between the numbers of so-called spindle cells, which are far more prolific in the right than in the left frontal lobes. At the level of biochemical pathways, differences were found between the projections in the two hemispheres of dopamine and norepinephrine, which are among the main chemicals (neurotransmitters and neuromodulators) playing a central role in signal transmission in the brain: slightly more dopamine pathways on the left and slightly more norepinephrine (noradrenergic) pathways on the right. Finally, at the molecular level left–right hippocampal asymmetries were found in the distribution of the microscopic subunits of the NMDA receptors. The NMDA receptors play an important role in memory and learning, since they enable signaling between neurons mediated by glutamate, one of the most prevalent neurotransmitters in the brain. As we already know, the hippocampi are the brain structures particularly important in memory. *And without exception, we share all these hemispheric differences with other mammalian species.*

So what was hoped to be a source of explanation soon became a source of confusion. To turn the argument around, if you believe that different functions require different structures, then different structures imply different functions. But in a chimpanzee or a gorilla, let alone in a rat or a mouse, the difference in function cannot be understood as the difference between language and nonlanguage. Much as we respect the mental powers of our fellow mammalians and are aware of their diverse and sometimes intricate communication methods (wolves' howls and whales' songs, to name two), they are *all* nonlanguage!

Of course, the inveterate romantics stubbornly subscribing to the notion of animal language may not buy my argument. They may even turn it around and conclude that the existence of hemispheric differences in other species in fact amounts to evidence in favor of animal language. If so, they better be prepared to carry this argument very far. Recent work by Alberto Pascual and his colleagues has demonstrated the existence of brain asymmetries in a certain group of . . . fruit flies. These asymmetries give them a distinct advantage over the less-fortunate fruit flies with symmetric brains. While both kinds of fruit flies could form short-term memories, only flies with asymmetric brains could effectively form long-term memories. So brain asymmetry appears to be a very basic and phylogenetically ancient device predating the emergence of language by millions and millions of years. . . . Unless, of course, you believe in fruit fly language!

It was becoming increasingly clear that a new conceptual framework was needed, a paradigm shift in our thinking about the duality of the brain. The search for a new paradigm was finally underway, made inevitable by the flood of new findings. The dominant role of the left hemisphere for language was not in dispute, but the centrality of this fact in explaining the differences between the two hemispheres was. It looked increasingly like the different roles of the two hemispheres in language was but a special, derivative case of some more fundamental, yet to be discovered difference, one that could be meaningfully observed and understood both in humans and in animals. *What is this difference?*

As is often the case, when rigorous science is at a loss, loose metaphors fill the void. A number of such high-profile metaphors were floated. The left hemisphere was declared "sequential," and the right hemisphere "simultaneous." The left hemisphere was declared "analytic," and the right hemisphere was declared "holistic." The problem with these metaphors was precisely that they were only metaphors, instruments of poetry and not of science. It was next to impossible to test them through clear experiments, or to use Karl Popper's famous term to *falsify* them. But in science propositions that cannot be proven false even potentially, even in principle, that are so elastic as to put even the Delphic oracle to shame, can neither be accepted as true. And so these sweeping metaphors made greater headway in the popular press than in serious scientific discourse. The search had to go on.

A New Paradigm: The Novel and the Familiar

My own interest in the brain's duality culminated in a theory very different from those dominating the mainstream neuropsychology at the time. It focused on the difference between the old and the new. I assumed that in order to understand how the hemispheres differ, one needs to take a dynamic approach and to look at the *processes* in the brain rather than at the static constants. Our mental life is always in flux, and the operative word is *learning.* By learning I mean far more than classroom exercises; I mean the process of mastery of the world outside—and of the world within—in all its rich and numerous manifestations. It usually does not happen instantaneously, as a miraculous, deus ex machina epiphany, a precipitous switch from total ignorance to perfect knowledge. It is usually a process.

I posited that the two hemispheres play different but complementary roles in this universal process, and that the two hemispheres differ in their relationship to novelty and familiarity. The right hemisphere is the novelty hemisphere, the daring hemisphere, the explorer of the unknown and the uncharted. The left hemisphere is the repository of compressed knowledge, of stable pattern-recognition devices that enable the organism to deal efficiently and effectively with familiar situations of mental routines.

The novelty-routinization idea came to me many years ago in the late 1960s, when I was a fledgling neuropsychology student working with Aleksandr Luria at the Burdenko Institute of Neurosurgery in Moscow. It was there that I found out that in children the effects of left-hemisphere damage were far less devastating than in adults, and by contrast the effects of right-hemisphere damage were much more devastating than in adults. I felt that if these observations were true, then their implications were quite sweeping. They suggested a broad transfer of cognitive control from the right to the left hemisphere in the course of cognitive development, and possibly through the whole lifespan.

But these observations were, at the time, merely a collection of anecdotal clinical impressions—a pretty shaky foundation for a grand theory. More systematic evidence was obviously needed to support or refute it. As is often the case in science, a provocative observation becomes a starting point for a systematic research program. But it had to be put on hold for a few years, while I was plotting and then executing my escape from the Soviet Union and getting settled in my new home, New York.

Exactly what makes the right hemisphere better suited to deal with novelty and the left hemisphere to be a repository of mental routines must obviously have something to do with the subtle differences in their wiring. Based on the new evidence that was just beginning to pour in at the time, I concluded that two such subtle but far-reaching differences between the wiring of the two hemispheres exist in the brain.

The first difference relates to the way in which the overall hemispheric surface is allocated to different types of cortex. In the right hemisphere it seems to favor the *heteromodal association cortex*; but in the left hemisphere it seems to favor the *modality-specific association cortex*. Both types of cortex are engaged in complex information processing, but in different ways. The modality-specific cortex is restricted to processing information arriving via a particular sensory system, visual, auditory, or tactile, and separate areas exist in the cortex for each of these sensory systems. Modality-specific cortex dismantles the world around us into separate representations. By way of analogy, think of an object in a three-dimensional space projected onto the *x*, *y*, and *z* coordinates, which generates three partial representations: This is what the modality-specific association cortex does with the incoming information. By contrast, the heteromodal association cortex is in charge of integrating the information arriving via different sensory channels, for putting the synthetic picture of the multimedia world around us back together.

The second difference relates to the way in which various cortical regions are connected in the two hemispheres. The left hemisphere seems to favor more local connections between adjacent cortical regions. By contrast, the right hemisphere seems to favor more far-flung connections between distant cortical regions. The connectivity of the left hemisphere is more like a fleet of taxicabs: You use them to go from one end of town to the other, but not to go from one end of the continent to the other. The connectivity of the right hemisphere is more like a fleet of airplanes: You use them to go from one end of the continent to the other. The story of the spindle cells is particularly interesting in the context of this idea. You may recall from the previous chapter that spindle cells relay information across very far-flung, distant brain regions. And true to the idea of hemispheric connectivity developed in this chapter, spindle cells are far more prevalent in the right hemisphere than in the left hemisphere throughout all the species studied.

Clearly, in its "canonical" form the novelty-routinization theory applies mostly to the right-handers among us. Most of the research on hemispheric specialization involves right-handed subjects; therefore, the dynamics of hemispheric interaction in left-handers must for now remain a matter of conjecture. Hemispheric specialization is less well articulated in left-handers and ambidextrous individuals, and the two hemispheres are functionally more alike. Interestingly, they are also structurally more alike, with the Yakovlevian torque reduced or altogether absent. In about 60 to 70 percent of left-handers, the profile of hemispheric specialization approximates the one seen in right-handers. It is reasonable to assume that the right hemisphere is in charge of novelty and the left hemisphere is in charge of cognitive routines, and that the right-to-left hemispheric shift occurs in these people. In about 30 to 40 percent of left-handers, the profile of hemispheric specialization roughly approximates the inverse of the one seen in right-handers. It may be not too far-fetched to assume that in these people the left hemisphere is in charge of novelty and the right hemisphere is in charge of cognitive routines, and that the hemispheric shift is left-to-right rather than right-to-left. But while the exact directionality of hemispheric dynamics over time may vary depending on the handedness, the general premise, that one hemisphere is in charge of novelty and the other is in charge of routinization, remains valid.

The ramifications of the novelty-routinization hypothesis were far-reaching, and they spelled a radical departure from the ways the roles of the two hemispheres were considered before. Far from assigning a fixed repertoire of roles to each hemisphere, the novelty-routinization hypothesis predicted an ongoing change in the nature of interactions between the two sides of the brain. What is novel today will become familiar tomorrow, in a week, or in a year. Appropriate patterns will be formed, and the problem that today can be solved only through strenuous, exacting mental effort will in due time be solved through near-instantaneous pattern recognition. The novelty-routinization hypothesis also challenged another unspoken tenet of traditional neuropsychology: that the functional organization of all the humans brains is exactly alike. But what is novel for one person is familiar to another person. Therefore, the novelty-routinization hypothesis implied a higher degree of individual differences in the ways our brains function than ever before imagined.

My idea could be wrong; in fact, in the beginning I only half-believed it myself, suspecting that it may be more elegant than true. But it certainly met Popper's falsifiability criterion (falsifiability is the sine qua non of any serious science), which set it apart from many other theories of hemispheric specialization aired at the time. The falsifiable prediction that followed from my idea was straightforward and unambiguous. If it did not hold up, the whole theory would fall apart like a house of cards. Any process of forming a new pattern—whether it was descriptive (learning a new concept) or prescriptive (learning how to solve a new class of problems)—had to engage first the right hemisphere and then the left hemisphere. There had to be a gradual shift of the "mental center of gravity," and the direction of this shift had to be highly predictable, regular, and unidirectional: from right to left.

Another appealing feature of my idea—and a very important one given the sudden flood of evidence that nonhuman brains are also asymmetric—was that the distinction between novelty and familiarity is meaningful not only for humans but also for any creature capable of learning. Animals also form patterns that allow them to navigate their world through the mechanism of pattern recognition. My bullmastiff Brit responds to a familiar command ("sit," "come," "down," and "no") given by any member of my office staff, despite the fact that he had learned them from my voice. And he knows not to enter certain areas (like the kitchen or the bathroom) in any apartment or office.

Brit has also developed an uncanny ability to recognize doormen, because in midtown Manhattan, where we live, many doormen carry biscuits for the neighborhood dogs. So now Brit stops in front of any doorman, even when he encounters a particular doorman and a particular building entrance for the first time. He squats in front of the good man, fixes an eagerly expectant gaze on him, and refuses to budge, awaiting his biscuit. He reserves this behavior for doormen and doormen alone. For the life of me, I have yet to figure out exactly how he tells doormen from the rest of humanity, but he does, this being an example par excellence of nontrivial, self-taught pattern recognition in a member of a nonhuman species.

My examples of bona fide learned pattern recognition revolve around canines because this is one mammalian species with which I have had extensive lifelong experience. But similar examples can undoubtedly be found for numerous other species, because the distinction between novelty and routinization is meaningful for them as well. Therefore, the novelty-routinization hypothesis can, at least in principle, serve as the basis for unraveling the mystery of the brain's duality across the mammalian evolution. Remote as it may sound in the context of our quest for understanding our own humanity, the puzzle of how to fit the brain's duality into an evolutionary context has been among the most fundamental and most intractable challenges facing cognitive neuroscience. It appears that the novelty-routinization hypothesis brought us closer to solving the puzzle than any hemispheric theory based on the language–nonlanguage distinction could aspire to do even in principle.

11

BRAIN DUALITY IN ACTION

All Patterns to the Left, Please

As it continued to jell in my head, I was both captivated by the novelty-routinization idea and slightly intimidated by its audacity. I felt that I needed a reality check and found it in the persona of an older colleague, my close friend, the neuropsychologist Louis Costa. Together, we formulated several tests of the hypothesis. We wanted to make sure that it could withstand the rigors of falsification, Popper-style.

Nothing is more exhilarating in science than to make a well-reasoned but risky prediction at loggerheads with the established precepts of the time and to see it confirmed—to stick your intellectual neck all the way out, so to speak, and to get away with it. You hope against hope, while cautioning yourself to only half-believe your own idea, lest the disappointment will be too bitter if it flops. Public recognition and accolades are sweet to a scientist, but nothing compares, at least for me, with the quiet, very private pride of testing the power of your own intellectual mettle and passing the test.

Our early conclusions were based on two kinds of observations. The first line of evidence involved comparing the effects of left-sided and right-sided brain damage on mental functions in patients. The second line of evidence involved the studies of healthy individuals using the experimental devices available at the time: a light box called a tachistoscope for visual processes and a set of earphones for auditory processes. These methods were extremely useful in the 1960s, '70s, and '80s, but they were crude and imprecise. In retrospect, I sometimes refer to them as "Paleolithic," a prehistory of cognitive neuroscience rather than its true history.

To our amazement and delight, our predictions were holding up. Not satisfied, we kept coming up with additional tests of the hypothesis, and they too were holding up. With science increasingly becoming a group effort today, too often run like a high-tech assembly line in an almost corporate spirit, its purest joy, the joy of clarity of thought, the unique kind of satisfaction found in solitary cerebral pursuits, is often missing. But this was that kind of moment, one of the high points of my career. We finally felt confident enough to publish our theory in 1981, in a journal article with an arcane title probably reflecting my Russian accent at the time, "Hemisphere Differences in the Acquisition and Use of Descriptive Systems."

The right-to-left shift of mental control looked increasingly like a universal phenomenon, capturing the essence of every learning process on every time scale, from hours to years. An individual faced with a truly novel situation or problem tackles it mostly with the right hemisphere. But once the situation becomes familiar and is mastered, the dominant role of the left hemisphere becomes evident. It looked like the empowering patterns capturing the essence of the situations (or rather the whole class of similar situations) were, once formed, stored in the left hemisphere.

The emerging evidence challenged even the most sacrosanct tenets of neuropsychology; it did not seem to matter whether the task involved language. What did matter was only whether the task was novel or familiar. A verbal task, but with an unusual twist (like figuring out which letters of the alphabet are present in a word, or matching verbs to nouns) engaged the right hemisphere more than the left hemisphere, even though according to the old precepts any task involving language should engage the left hemisphere. But as the "twisted" task was becoming more familiar, the left hemisphere was getting increasingly involved. By contrast, a verbal task more closely approximating the way we use language in everyday life activated the left hemisphere from the beginning.

Likewise, a visuospatial task of a familiar nature (like recognizing familiar faces) engaged mostly the left hemisphere—even though according to the old precepts any visuospatial task, including facial recognition, should predominantly engage the right hemisphere. By contrast, comparing the photographs of unfamiliar faces activated mostly the right hemisphere. And so on.

More recently, powerful methods of functional neuroimaging became available and they revolutionized brain research. All of a sudden, the traditionally low-tech, paper-and-pencil based neuropsychology added to its vocabulary exquisitely high-tech terms like PET, fMRI, and SPECT and sophisticated forms of electroencephalography (like MEG, or the recording of "gamma-frequency" associated with complex decision-making). All these methods are based on different physical principles, but they all give us a direct window into the activity of a working brain in action. The information obtained by these methods is "macroscopic," a bird's-eye view of a working brain, rather than a cerebral magnifying glass. It does not tell us about the activity of an individual neuron, or even of an individual neuronal circuit. But despite all their limitations, these methods do tell us which constellations of brain regions and which structures become active under what conditions.

The new functional neuroimaging methods allow a much more direct and precise look at the brain dynamics and at the changes in brain activity over time. In the last few years a wealth of additional information has become available, further clarifying the roles of the two hemispheres in learning. The new methods support the conclusion that the "cognitive center of gravity transfer" from the right side of the brain to the left side of the brain is a universal rule inviolate across various cognitive tasks, from verbal to visuospatial, and across various time scales, from hours to decades.

The transfer could be demonstrated within a single experiment lasting a few hours in a lab, when subjects were asked to learn previously unfamiliar tasks of various kinds. Invariably and regardless of the nature of the task, the right hemisphere was dominant in naive individuals at early stages of acquiring a cognitive skill, but with increased task mastery the left hemisphere was taking over. This is illustrated in a gamma-frequency EEG study by Japanese neuroscientists using an ambiguous novel task prompted by our work on actor-centered decision-making.

The right-to-left transfer could also be demonstrated for various real-life professional skills, which take years to acquire. Novices performing the tasks requiring such skills showed clear

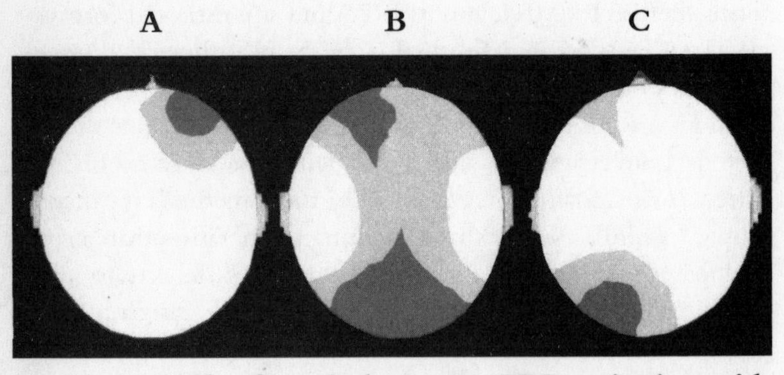

FIGURE 14. **Shift of cortical gamma EEG activation with task familiarization.** *The darker the shading, the greater the activation level. (A) Initial exposure to the task—the right hemisphere is mostly active. (B) Halfway through experiment—both hemispheres are active posteriorly, but frontal lobes are active mostly on the left. (C) Toward the end of experiment—the left hemisphere is mostly active. Adapted with permission from Kamiya et al. (2002).*

right-hemisphere activation. But skilled professionals showed distinct left-hemisphere activation while performing the same tasks. Music is a good example. When musically untrained individuals (like most of us) were asked to recognize melodies, the right hemisphere did a better job and was particularly active. But in professionally trained musicians the opposite was true: The left hemisphere did a better job and was particularly active.

Even the lateralization of language, the holy grail of traditional neuropsychology, was not what it was believed to be. It was not true that language was the monopoly of the left hemisphere from the beginning. The right hemisphere turned out to play an important role in language acquisition in little children. This was obvious from the studies of normal children using various experimental methods. But particularly compelling evidence arose from the effects of brain damage on language. Consistent with my earlier observations in Russia, in children damage to the right hemisphere significantly interferes with subsequent language development. By contrast, in adults damage to the right hemisphere usually does not disrupt language, but damage to the left hemisphere does. But if the verbal task is highly novel or unusual, then the right hemisphere will become involved even in adults. Mark Jung-Beeman and his colleagues demonstrated this in an elegant experiment using problems like the following: "What single word will produce compound words if combined with *pine, crab*, and *sauce*?" (Solution: *apple*, resulting in *pineapple, crabapple,* and *applesauce*). When the solution of such problems involved "Eureka-like" insight, right-hemispheric activation was noted on fMRI and EEG.

Language acquisition as pattern learning begins at a very early stage, with learning the sounds of one's native language: *phonological learning.* I am a native Russian speaker and no matter how good my English is (good enough for me to be on my third book in this adopted language of mine), my language processes and the underlying brain processes are somewhat different from those of a native speaker.

Our country is still a land of immigrants. While many of them never learn the language of their new homeland beyond utilitarian, stripped-down, bare-bone English, there are also those who attain an astounding level of linguistic proficiency, even virtuosity in their adopted tongue and become prolific public speakers and authors. One thinks immediately of Vladimir Nabokov and Joseph Conrad. But also more recently of Henry Kissinger, Eli Wiesel, and George Soros, to name a few. They are all spectacularly articulate in English, probably more so, at a certain stage of their lives, than in their respective native languages. Yet their command of English is different from that of a native speaker. Not worse, possibly even better in many instances, but different. And these differences go much deeper than their identifiably Middle- and East-European accents, which usually do not completely disappear in people first exposed to a second language when they are past their teens.[1]

Some of these differences are transparent, predictable on commonsense grounds and thus unsurprising. Usually one acquires different aspects of the lexicon at different ages. This means that if you were first exposed to your adopted language past a certain age, your grasp of certain aspects of your vocabulary will remain tenuous for the rest of your life. The paradox is often that the simpler the words, the more this rule applies. A highly articulate individual will be able to speak eloquently about the most rarefied strata of science, philosophy, or politics, yet will be comically unsure about the names of simple household gadgets, plants, or animals in his or her second language. I find myself in such ridiculous, comical yet linguistically humbling situations all the time.

---

[1]Amusingly and incongruously, one of the heaviest accents in a superarticulate multilingual belonged, in my experience, to Roman Jakobson (1896–1982), a Jewish-Russian émigré turned Harvard professor and one of the most prominent American linguists, renowned to this day for his studies in the phonological structure of languages.

A more elusive, but generally more profound, difference between a native speaker and a very proficient nonnative language speaker involves attention, the relationship between *hearing* and *listening.* No matter how articulate I may sound and how adroitly I grasp the information conveyed to me in English, my English is invariably more effortful than that of my culturally and intellectually equal native speaker. It taxes my attention to a far greater extent. This is true even if the *perceived* quality of my English is indistinguishable from native, as it very often is, if one ignores the accent. To put it plainly, in English I have to listen in order to hear; in Russian I can hear without listening. Any interaction in a second language forever requires extra mental resources, is more effortful. This may be one of the reasons why in cognitive decline and early dementia bilingual individuals frequently lose the command of their second language first and revert to the native language despite the decades of conversing predominantly in their second language, as Stalin infamously did.

Does the brain process a native language and a nonnative language differently? New research shows that indeed it does. For many years, decades even, scientists operated on the assumption that the brain mechanisms of language are uniform and modular, that language inhabits in everybody the same parts of the brain, several predictable regions of the left hemisphere. But thanks to a recent surge of groundbreaking insights and findings that transformed cognitive neuroscience, we know today that the brain machinery of language is far from static. Different stages and different degrees of language development rely on constellations of different brain regions. As we already know, the right hemisphere plays an unexpectedly vital part at early stages of language development in children. The role of the right hemisphere in language recedes gradually with age and becomes relatively limited in the adult brain. This has been shown by comparing normal children of various ages and also by examining the effects of hemispheric damage sustained at different ages. Such a pattern of brain dynamics in native language development is in keeping with the general principle of brain organization: The right hemisphere is in charge of dealing with novel information, and the left hemisphere is in charge of dealing with well-established cognitive skills.

The brain dynamics of a second or third language are even more complex than that of the first language. A second language is by definition novel. Yet it is not entirely novel, since different languages have a lot in common, and the acquisition of a second or third language builds on the already well-entrenched first language. Recent studies of bilingual individuals using functional neuroimaging have shown that the brain regions activated in the first and second languages are not identical, even though a considerable overlap exists. The brain dynamics of the first language in an adult bilingual are by and large restricted to the left hemisphere. By contrast, the brain dynamics of the second language usually involve both the left and right hemispheres. The bulk of evidence to this effect comes from functional neuroimaging studies in neurologically healthy bilingual individuals. (But anecdotal reports also exist of bilingual individuals reverting to their first language after decades of using the second language following a right-hemispheric stroke.)

So it appears that when language is still a relatively novel cognitive device (as is the case with the first language in a child and with a second language in an adult), the right hemisphere plays a critical role in its acquisition. But as the language becomes fully acquired, it is gradually monopolized by the left hemisphere. As we established earlier, language is a system of generic patterns, and these patterns are stored in the left hemisphere as they are being formed in the person's brain.

Kinds of Patterns

And so are other generic patterns. With all due reverence toward language and its role in human cognition, our mental world is replete with other nonverbal pattern-recognition processes, which are guided by generic memories that are relatively independent of language. As we have already established, in navigating even the most mundane everyday situations, we critically depend on our ability to instantly recognize unique exemplars as members of familiar categories, even if we have never encountered these particular exemplars before. But if you have never seen these things before, how do you know what they are? Pattern recognition to the rescue! Each of these representations is nothing other than a neural network in your brain with attractor properties (we talked about attractors earlier): It will be activated by a whole class of sensory inputs, corresponding to a whole class of similar things. We exercise this ability practically every moment of our lives. When you see a new car model, you know nonetheless that it is a car and not a palm tree. When you walk through the aisles of a department store, you don't have to be told which item is a shirt and which one is a pair of shoes, and on and on and on. Without this ability we would be lost in the forest of strange, befuddling objects and would have to learn the meaning of each of these objects from scratch.

Certain kinds of brain damage disrupt this precious ability, producing a condition known as "associative agnosia." Particularly germane to our discussion is that associative agnosia may be caused by damage to the left hemisphere or both hemispheres, but not by damage to the right hemisphere alone. So the left hemisphere is the seat of all kinds of generic patterns, both those related to language and those unrelated to language. Assuming that you are right-handed, this means that the neural network representing the concept of "chair" in your head, while probably widely distributed, inhabits mostly the occipital, temporal, and parietal lobes of your left hemisphere.

Not all generic patterns are descriptive. Some are prescriptive, and these latter generic patterns are also stored in the left hemisphere. We discussed this in the previous chapter, but let me enlarge on the subject. Not only do we know what various objects are, we also know what to do with them. We know how to hold a spoon, a comb, a pen. The hand positions for holding each of these objects are all different and we don't confuse them. We know how to tie shoe laces and a tie, how to push buttons through buttonholes, how to handle a hammer and a pair of scissors. The movements associated with all these objects are also different and we don't confuse them either.

Furthermore, as with the Art Deco chair, we don't have to learn these motor skills for each individual object separately. Once you know how to manipulate scissors, you can do it with any pair of scissors; once you know how to tie a tie, you can pretty much tie any tie, regardless of its length or width. This is why such motor skills are also generic. Certain kinds of brain damage may impair these skills, producing a condition known as *ideational apraxia*. And again, ideational apraxia may be caused by damage to the left hemisphere or both hemispheres, but not by damage to the right hemisphere alone. So the prescriptive generic patterns are also stored in the left hemisphere, whether they are related to language or not.

To sum it up, the left hemisphere is in charge of most of the processes based on pattern recognition, both those that involve language and those that do not. Damage to the left hemisphere disrupts these abilities, producing both deficits of language (aphasias) and deficits of nonverbal pattern recognition and pattern use (agnosias and apraxias).

By contrast, the right hemisphere plays a particularly crucial role at the early life stages, while the arsenal of ready-to-use patterns is still limited. This has finally been acknowledged by developmental neuropsychologists. For many years the tacit assumption ruled that all the disorders of learning and early cognitive development stem from the malfunction of the left hemisphere. But in recent years, a host of conditions stemming from early dysfunction of the *right* hemisphere has been described: the so-called nonverbal learning disabilities, Asperger's syndrome, and others. The Canadian neuropsychologist Byron Rourke in particular has contributed much to our understanding of right-hemispheric dysfunction in various developmental disorders.

Some of the symptoms caused by right-hemispheric dysfunction can be revealed only through neuropsychological tests. But other symptoms are apparent even to the naked eye of an uninitiated observer, and they speak volumes about the function of the right hemisphere by highlighting what is impaired when this hemisphere is damaged. People with right-hemispheric dysfunction usually eschew novel situations. They tend to cling to routines and be rigid, fearful, and resentful of any departure from well-entrenched scripts in any life circumstances.

These symptoms of right-hemispheric dysfunction can be quite dramatic, and they particularly affect social behavior. Some people are socially nimble, and others are socially awkward. What's more, social awkwardness is often present in impressively accomplished individuals: scientists, engineers, and software designers—the proverbial geeks. In small doses, such awkwardness may even be endearing, but once it reaches a certain level of severity, it becomes extremely disruptive to the person's very fabric of life. Today, we know that such clinically devastating social awkwardness is often produced by damage to the right hemisphere.

How so? The answer lies in the fact that certain types of situations never quite lend themselves to being forced into a finite number of patterns. To deal with them effectively, the individual must constantly improvise and rely on his or her "sense" of the situation rather than on slam-dunk pattern recognition. This means that certain types of decisions forever remain the purview of the right hemisphere. Social judgment and the ability to navigate interpersonal relationships seem to fall into this category. Social situations are just too diverse, too fluid, and too nuanced to lend themselves to codification through a finite number of templates.

What distinguishes a socially graceful individual from a socially awkward one is less the knowledge of social norms and more the ability to adhere to them smoothly, in a way that does not come across as contrived, labored, or artificial. We all know people who in social situations do everything right, "by the book," which is precisely why they come across as scripted martinets, almost like in pantomimed modern-dance caricatures, or highly stylized theatrical performance. Their behavior comes across as a sequence of awkward still shots, each corresponding to the "concept" of behavior rather than as natural behavior, devoid of nuance, fluidity, and grace. They try desperately to "fit in" but end up putting their social feet in their mouths at every turn and are shooed and rejected by their peers as social aliens if not social outcasts. In patients with right-hemispheric dysfunction these traits are often quite pronounced, and not just in children but also in adults.

As we move from childhood to adulthood, we accumulate various patterns enabling us to deal with new situations as if they were familiar. Once formed, these ready-to-use patterns are stored mostly in the left hemisphere, and as their repertoire grows, the individual relies increasingly on the left side of the brain. The overall locus of cognitive control, "the center of mental gravity," gradually shifts away from the right hemisphere to the left hemisphere. This is obviously a process and not a precipitous jump, and this process is different for different cognitive skills. So, to be precise, we are not talking about a single process, one grand right-to-left shift, but rather about a myriad of such processes unfolding in parallel, on different time scales and at different speeds. But they all represent a single fundamental phenomenon. The right-to-left shift of the locus of cognitive control is a fundamental cycle of higher-order mental processes, just as reflex is a fundamental unit at a more elementary level of learning.

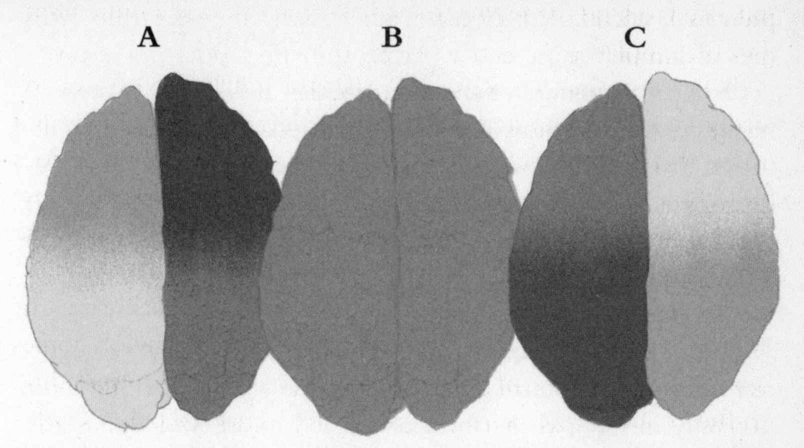

FIGURE 15. **The roles of the two hemispheres in cognitive learning.** *The darker the shading, the greater the level of involvement. (A) The right hemisphere is dominant when you are confronted with a novel cognitive challenge. (B) Both hemispheres are equally involved at an intermediate learning stage. (C) The left hemisphere is in charge of "cognitive autopilot" in exercising well-developed mental skills. To some extent, the role of the prefrontal cortex (top part of images) also decreases in well-established mental skills.*

A Lifelong Shift

Now let's consider a lifelong perspective. In most studies of the changes in brain function with development, the implicit question is, "How do children differ from adults?" But recently, with the growing interest in aging, the question has been extended to "How do younger adults differ from older adults?"

Over the last decade, this question was asked in a number of functional neuroimaging studies using PET and fMRI. The patterns of brain activity were compared in adults at different stages in their lives. The findings showed an ongoing progression of the right-to-left shift of the "center of cognitive gravity" throughout the life span. In younger adults, a decidedly greater extent of activation is present in the right prefrontal cortex than in the left prefrontal cortex. But in older adults, the left prefrontal cortex becomes much more active. Again the effect does not seem to depend on the nature of the task, whether it is verbal (like word recognition) or visuospatial (like facial recognition). The right-to-left shift of the center of cognitive gravity seems to be a lifelong phenomenon extending from childhood through middle age through advanced age. This idea, first proposed by two prescient friends of mine, Jason Brown and Joseph Jaffe, is finding increasing empirical support.

So it appears that the right-to-left shift in the locus of cognitive control is a fundamental cycle of the journey of our mind not only in our passage from childhood to adulthood, but also throughout our whole life span. In the beginning of this chapter, we discussed how these changes are studied in the laboratory. But now we know that similar changes occur on the scale of human life. Contrary to previously well-entrenched beliefs, *the right hemisphere is the dominant hemisphere at early stages of life.* But as we move through the life span it gradually loses ground to the left hemisphere, as the latter accumulates an ever-increasing "library" of efficient pattern-recognition devices in the form of neural attractors. The right hemisphere is of foremost importance in our youth, the season of daring, of navigating uncharted waters. The left hemisphere is of foremost importance in mature years, the season of wisdom, of seeing new things through the prism of vast past experience.

How can we understand the differences in knowledge representation in the two hemispheres that account for their different roles at various stages of learning? As I am writing this book, these differences are the focus of intense research both with functional neuroimaging and with computational methods. But for now, the more scientifically minded readers of this book may find the following analogy helpful. The analogy involves descriptive statistics, the simplest way of representing large data sets even before any elaborate analysis ("inferential statistics") is conducted. In descriptive statistics, the same set of data can be represented in two different ways: as group data and as a cloud of individual data points. The first representation is a grand average that captures the essence of the totality of all previous experiences, but in which the details, the specifics, are lost. The second representation is a library of specific experiences, but without the ability to extract the essential generalities.

Group data are represented by means and standard deviations. By contrast, individual data points are represented by scatter-plot diagrams. When new information arrives, the two respective representations will be updated in two very different ways. The group data will have to be recalculated every time such new information is received, resulting in a new mean and a new standard deviation. By contrast, the scatter-plot diagram will be updated by merely adding individual new data points.

Think of the right hemisphere as representing the organism's cumulative knowledge through some cortical means and standard deviations of sorts, as the "grand means" of all prior experiences, but with the loss of details. Think of the left hemisphere

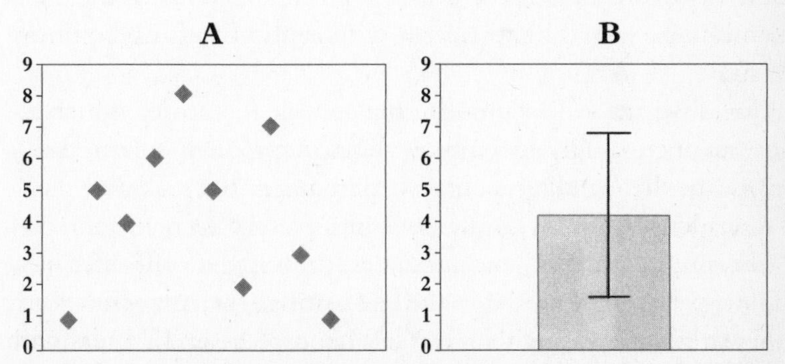

FIGURE 16. **How knowledge is represented in the two hemispheres.** *(A) Scatter plot (each data point capturing the specific properties of a narrow class of situations)—this is how knowledge is represented in the left hemisphere. (B) Means and standard deviation (very coarse averaging across all situations)—this is how knowledge is represented in the right hemisphere. This figure is a heuristic metaphor rather than a literal depiction.*

as a cortical scatter-plot diagram of sorts, as a library of relatively specific representations, each corresponding to a relatively narrow class of similar situations.

Suppose now that the organism encounters a new cognitive challenge. If it resonates with at least one of the specific representations (attractors) contained in the left hemisphere, the cognitive challenge is recognized as familiar and is dealt with according to previously acquired experience specific to that type of situation. But if no such resonance occurs, the cognitive challenge at hand is recognized as truly novel. Since it does not correspond to any situation-specific knowledge at the person's disposal, the only way to approach the situation is through the default "averaged" information contained in the right hemisphere.

Suppose, for instance, a jar with a jelly-like substance turns up in your kitchen. If your left hemisphere recognizes it as fruit jam, you may decide to eat it. If your left hemisphere recognizes it as liquid soap, you may put it in the dishwasher. But if you fail to recognize it as anything familiar at all, if in other words you don't know what it is, the default option contained in the right hemisphere would be to treat it with caution and maybe throw it away.

So as we age, we accumulate the generic memories, which allow us increasingly to employ shortcut problem-solving skills to escape the grinding mental work required to crack new mental challenges, and to condense it into pattern recognition. Our "library of patterns" accumulates throughout our lifetimes. These patterns are stored in the left hemisphere. As a result, with age we rely increasingly on our left hemisphere and decreasingly on our right hemisphere. As we accumulate mental patterns, the ratio of hemispheric use, so to speak, shifts inexorably from right to left. This, in turn, is fraught with its own set of profound consequences for the brain and for the resistance of the two hemispheres to the effects of aging. We will discuss this later in the book, in chapter thirteen.

But meanwhile it is time to examine the relationship between novelty, cerebral hemispheres, and the frontal lobes. The right hemisphere is not the only part of the brain that is important for dealing with cognitive novelty. We know from the previous chapter that the frontal lobes also play a critical role in this regard. Functional neuroimaging studies have shown that the frontal lobes are particularly active when the organism is faced with novel challenges. As tasks become familiar, autonomous, and effortless, the role of the prefrontal cortex diminishes.

Not surprisingly, creativity depends on frontal-lobe function as well. Ingegerd Carlsson and her colleagues studied prefrontal regional cerebral blood flow (rCBF) in people with low and high creativity (where creativity was measured with a special psychological test). Resting frontal rCBF levels were higher in the high-creativity group. When a cognitive challenge was introduced, the high-creativity group showed bilateral frontal activation, and the low-creativity group showed only left frontal activation. So it appears that both right and left frontal lobes participate in problem-solving in highly creative individuals, but in less creative individuals only the left prefrontal cortex participates. A similar study suggested that faced with a task requiring ingenuity, creative people rise to the occasion with an increased right-hemisphere activity. Right-sided activation is particularly pronounced in the frontal lobes. By contrast, the less-creative people remain at the mercy of the left hemisphere, with the right hemisphere being relatively inactive.

Since the transfer of the "center of cognitive gravity" from the right hemisphere to the left hemisphere seems to be a universal phenomenon of moving through life, the following question is in order: Does this mean that it occurs in all people with clockwork uniformity, or is there some room for individual differences? Knowing what we already learned about the brain and cognition, one might expect the latter.

Indeed, in some people creativity is a lifelong trait, undiminished by aging. Are their brains wired differently, and how?

Consider the following mental experiment. Suppose we had a psychological test at our disposal allowing us to measure wisdom. Suppose that with the help of this test we could identify wise and not-so-wise people. Suppose, further, that we faced our subjects with a problem whose solution required wisdom. What would be the differences between brain-activation patterns in the wise and not-so-wise people? I believe that a particularly strong activation of the left prefrontal regions would emerge as the hallmark of wisdom. And those among us who attained wisdom with age while retaining the gift of creativity will exhibit a particularly strong activation of both right and left prefrontal regions.

By better understanding the workings of the two cerebral hemispheres we come a step closer to solving the puzzle of some of the most mysterious aspects of cognition. But cognition does not operate by itself, as if in a dispassionate, emotionally neutral vacuum. Quite the contrary, cognition and emotion are intertwined, and this union also involves the two cerebral hemispheres. This will be the subject of the next chapter.

language. Since language is such an important and all-embracing skill, its loss cannot remain unnoticed by the patient and it becomes the source of intense distress. By contrast, the functions of the right hemisphere are more elusive, less available to introspection. Patients are usually less aware of the loss of these functions and are therefore less perturbed by their loss. The impression of nonchalance when there is every reason to be depressed could be misinterpreted as euphoria, as the reasoning went.

Indeed, a patient with right-hemispheric damage often displays astounding unawareness of the deficit, a phenomenon known as *anosognosia*. The blissful aura of equanimity projected by these patients stands in stark contradiction with the sad reality of the catastrophic brain damage many of them suffered.

Anosognosia often takes the form of "left hemineglect," a condition that occurs when the brain fails to properly register and process information coming from the left half of the outside world. The condition is possible because the sensory pathways carrying information about the outside world to the brain are mostly crossed: Information about the left-hand half of the world is sent to the right hemisphere and information about the right-hand half of the world is sent to the left hemisphere. When a lesion affects the left side of the brain, the patient usually discovers the consequent handicap quite easily and learns how to compensate for it. But when a lesion affects the right side of the brain, the patient often remains unaware of the consequences and fails to compensate, and the left hemineglect becomes severe and intractable.

Anosognosia sometimes takes rather surreal forms, when the failure to recognize a problem within results in fantastic accounts about the world outside, such as the nursing home patient discussed earlier who was unable to find his steak on the cafeteria tray and blamed this on a conspiracy of nurses. But left hemineglect and hemiinattention are not limited to the visual sensations. The tactile sensations may also be affected, producing the so-called "alien hand" phenomenon. A stroke patient afflicted with this condition will disown the left side of his own body as belonging to another person, will ad lib a bizarre story explaining what the "alien" hand is doing next to him, and will not be in the least concerned about his own neurological condition.

By contrast, a patient with aphasia (language impairment) caused by a left-hemispheric stroke is often acutely aware of his handicap and is tormented by it, frightened and tearful. This has frequently led to the conjecture that depression in such patients is a reaction to their cognitive loss.

But further research has shown that there is more to the connection between hemispheres and affect than the differences in the degree of awareness of deficit. A hemisphere is a big place, and linking certain symptoms to damage somewhere within a hemisphere is not enough. It is important to know *where exactly within the hemisphere* the offending damage is found. When this question was asked, it turned out that damage to the left frontal lobe is particularly likely to produce depression, more so than damage to any other part of the left hemisphere.

But therein lies a riddle. As we already know, frontal-lobe injury also causes anosognosia. A patient with significant damage to the left frontal regions is just not sufficiently aware of his or her deficit to be bothered by it. Therefore, linking depression to the awareness of deficit caused by a left frontal lesion amounts to a highly implausible proposition. On the other hand, damage to the right frontal regions often produces a much-more-than-blasé nonchalance, which could not be explained away simply by unawareness of deficit. Such lesions often produce mania or outright euphoria.

It has also been noted that at times damage to one or the other hemisphere produces emotionally charged behaviors so extreme that they cannot be explained by the degrees of awareness of deficit. Patients with left-hemispheric lesions sometimes engage in pathological crying and patients with right-hemispheric lesions occasionally engage in pathological laughter. So the side of hemispheric damage had to be linked to these changes of affect.

The next step was to study the relationship between emotional states and the two sides of the brain in normal people. This was first accomplished using EEG, which remained the mainstay of such research in the 1970s and 1980s. The advent of functional neuroimaging (PET, and fMRI) in the years that followed made possible an even more direct glimpse into the relationship between affect and the two sides of the brain. Much of this work has been pioneered by Richard Davidson and his colleagues.

The findings were quite intriguing. When normal subjects were shown film clips or other images containing pleasant information, the activation increased in the left hemisphere, particularly in the left prefrontal cortex. By contrast, when subjects were shown unpleasant or sad images, activation increased in the right hemisphere, again mostly in the right prefrontal cortex. A similar contrast was evident in a video game with financial implications. When the subjects stood to make money, a relatively greater activation of the left frontal lobe was recorded. But when the subjects stood to lose money, there was a relatively greater activation of the right frontal lobe. When the brain mechanisms of various spiritual experiences were studied, similar effects were found. Meditation leading to the immersion into a soothing, introspective frame of mind activated the left prefrontal cortex and decreased the right prefrontal activation. An increase of activation was found in the left frontal regions in meditating nuns, as well as a decrease in various regions of the right hemisphere.

Taken together, the studies of brain damage and the neuroimaging studies in normal people clearly indicated that the two hemispheres play rather direct, opposite roles in the experience and expression of emotions. The left hemisphere mediates positive emotions and the right hemisphere mediates negative emotions: truly the Yin and Yang in the brain.[2]

---

[2]Personal emotional experience and expression should not be confused with the ability to recognize and discriminate emotional expressions in other people.

The next step was to explore the individual differences in emotional styles. Davidson and his colleagues discovered that different emotional styles indeed exist and that they correspond to the predominance of left-hemispheric or right-hemispheric activation. Some people tend to have a positive, cheerful disposition and others are susceptible to depression. It turns out that their brain activation profiles differ in a stable, consistent way, revealing different electrophysiological traits. The left frontal areas tend to be more active in the happy-go-lucky types, and the right frontal areas are more active in the brooding types inclined toward depression. If, for some reason, the activation of the left frontal areas is impaired, sadness and depression set in. Similarly, the activation of the right frontal areas is associated with distinctly negative emotions, such as disgust or fear. Even such highly complex feelings as painfully negative emotional reactions to social exclusion are lateralized and involve the right frontal cortex. This was demonstrated by Naomi Eisenberger and her colleagues in a clever fMRI experiment where some of the subjects playing a virtual ball game were excluded from the game.

The differences in emotional styles and their link to the two hemispheres appear to be innate, or at least they appear very early in life. Left frontal activation was found to be particularly strong in cheerful ten-month-old infants and right frontal activation was particularly strong in similarly aged crybabies.

The hemispheric division of labor in regulating our emotional world is not limited to the neocortex. It involves the amygdala also. In healthy individuals the left amygdala is more active in response to positive stimuli than to negative stimuli. By contrast, anxious people tend to have an exaggerated activation in the right amydgala while viewing fearful and neutral faces;

The latter ability seems to depend mostly on the right hemisphere both for positive and negative emotions—probably due to the fact that this type of information processing does not easily lend itself to pattern recognition.

and depressed people exhibit reduced activation in the left amygdala. This evidence leads to the conclusion that two cohesive "emotion circuits" exist, each involving the frontal lobes and the amygdala in one or the other hemisphere. Indeed, it has been shown that the fronto-amygdaloid circuitry is involved in many decision-making processes associated with rewards, including even the process of selecting the most appealing items from the restaurant menu.

In certain psychiatric conditions, patients differ not only in their pattern of brain activation, but also in the outright size of their brain structures. Patients suffering from Generalized Anxiety Disorder often have a particularly large right amygdala. By contrast, patients who underwent surgical resection of the right amygdala to relieve intractable seizures (a procedure called anterotemporal lobectomy) lost the ability to appreciate facial expressions of fear.

Other brain structures are also involved in the regulation of emotions. These include the cingulate cortex (a ribbon of ancient cortex hugging the outer boundary of a massive bundle of pathways connecting the two hemispheres, the corpus callosum) and certain parts of the thalamus (a subcortical collection of nuclei projecting to various cortical regions). Little is known about the functional lateralization of these structures, but it is very likely that they parallel the division of labor between the left and right cortical hemispheres.

So it appears that the prefrontal cortex, the amygdala, the cingulate cortex, and possibly other structures operate in concert in mediating emotional experience and expression, and that they comprise two distinct, parallel systems of emotional control. On the left side of the brain this system mediates positive emotions, and on the right side of the brain it mediates negative emotions. Of course, most real-life experiences are complex. They are more likely to be bittersweet than purely sweet or purely bitter, like the Yin and the Yang, or the black-and-white symbol of the classic Balinese ornamental design. Therefore, in most real-life situations the two fronto-amygdaloid loops operate in concert, but their contributions to the emotional balance differ.

Convergence of the Themes

A careful reader of this book has probably noticed already that the quest for understanding the nature of hemispheric specialization took several parallel paths, unfolding without much cross talk or convergence. The first path was mostly concerned with cognition; it pursued the notion that the left hemisphere was the language hemisphere, and the right hemisphere was the visuospatial hemisphere. As we already know, this has been the main theme in neuropsychology for many years. The second path, of a more recent vintage, was mostly concerned with emotions, and it pursued the relationship between the two sides of the brain and the negative and positive affect.

These two strands of neuropsychology have never really managed to interweave. They have existed in mutual isolation, pursued by different crowds of neuroscientists, discussed at different scientific meetings, and written up in different scientific monographs. Amazing as this state of affairs may sound, it was not entirely surprising. There is no logical or empirical way to argue that an intrinsic relationship exists between language and positive affect, and between visuospatial functions and negative affect. Language is an emotionally neutral, or rather emotionally equipotential, tool. It contains in equal measure the means for encoding and expressing both positive and negative emotions. Likewise, visual imagery also lends itself equally well, or equally poorly, to rendering either type of emotions.

In formal scientific parlance, the language-visuospatial distinction and the positive–negative emotional distinction are *orthogonal*, irreducible to one another. So what does this mean? That their parallel affiliations with the two hemispheres are merely coincidental? Science has always thrived on the esthetic imperative of parsimony, the ability to relate a multitude of observations to a minimum number of underlying principles. The imperative of parsimony has been traditionally so widely embraced in scientific discourse that esthetic and explanatory considerations often comingled in an almost interchangeable way. A parsimonious theory has face value, is more believable, more compelling, and more likely to be accepted as containing a genuine explanation. By contrast, a theory lacking in parsimony is automatically suspect as to its true explanatory power. In a compelling scientific theory, various themes must converge.

By this standard, the coexistence of two or more orthogonal, seemingly coincidental principles of hemispheric specialization, should have been unsatisfactory and very disturbing to neuropsychologists and cognitive neuroscientists. Was it? Not necessarily. The whole field has become so fragmented that many scientists are concerned about the intellectual order only within their own relatively small niches and not across the niches. But it did concern me. The need for the parsimony capable of bringing the disparate strands of brain research together has been and remains my personal intellectual imperative.

The novelty-routinization theory of hemispheric specialization that we discussed in the previous chapter brings about the much-needed parsimony by tying together the cognitive and emotional aspects of hemispheric specialization in a way that earlier theories could not. This is so because an intrinsic link exists between cognitive routines and positive affect, and between novelty and negative affect. Here it how this works.

The left hemisphere is the hemisphere of cognitive routines. As we have already established, the brain is highly selective in admitting information into long-term store. In the normal brain, only such knowledge becomes routinized and committed to the long-term store in the left hemisphere as has been proven useful over a period of time. Useless information (like what you ate for lunch twenty years ago today) does not make it into the collection of pattern-recognition routines housed in the left hemisphere. So the content of left-hemispheric storage consists overwhelmingly of "useful" information, which by virtue of its utility is good for the organism.

By contrast, the right hemisphere deals with novelty. It steps in whenever the cognitive repertoire already at the organism's disposal fails to solve the problem at hand and when de novo exploration is required. Right-hemispheric involvement is triggered by a disparity between the organism's abilities and the organism's needs. The search for a novel solution is triggered by the dissatisfaction with the status quo, by a situation that is unsatisfactory, i.e., bad for the organism.

A look into brain biochemistry further highlights the close link between the cognitive and emotional aspects of hemispheric specialization. As we already know, the two cerebral hemispheres are not mirror images of each other either structurally or biochemically. Certain neurotransmitters are slightly more abundant in the right hemisphere than in the left hemisphere. This is particularly true for norepinephrine. Other neurotransmitters are slightly more abundant in the left hemisphere than in the right hemisphere. This is particularly true for dopamine.

Such biochemical asymmetries are of great consequence for both cognition and emotions. Animal experiments have shown that an increase in the brain dopamine levels triggers overlearned, stereotypic behaviors. Dopamine is associated with reward and the reinforcement of successful behaviors. Dopamine also plays a role in the experience of pleasure and in addiction. So dopamine appears to mediate positive emotions *and* cognitive routines. This makes perfect sense, since cognitive routines encode experiences that proved to be good (successful) in the past.

By contrast, animal experiments have shown that an increase in norepinephrine levels in the brain triggers restless exploratory behavior, incessant search for novelty. At the same time, abnormal levels of norepinephrine have been implicated in depression. So this neurotransmitter mediates both negative emotions *and* exploratory behaviors. This also makes perfect sense, since the organism's failure to meet its needs triggers both negative emotions and the search for new solutions. Interestingly, decreased levels of another neurotransmitter implicated in depression, serotonin, produce cognitive inflexibility, this pointing again to the unity of cognition and affect.

One might ask the following question: Are the hemispheric roles in emotions secondary to their roles in cognition? According to this scenario, the link between positive emotions and the left hemisphere is derived from the fact that left-hemispheric activation corresponds to intrinsically "good" situations (a good fit between the organism's needs and its abilities to satisfy these needs). Likewise, the link between negative emotions and the right hemisphere is derived from the fact that right-hemispheric activation corresponds to intrinsically "bad" situations (a mismatch between the organism's needs and its abilities to satisfy these needs). Or is it the other way around, the roles of the two hemispheres in opposite emotions being primary, and their roles in dealing with familiar as opposed to novel situations derivative?

This might be a bit of a chicken-or-egg question, not only unanswerable but also not that important in the end. But it is instructive that among all neocortical regions the prefrontal cortex becomes particularly active during emotional experiences: The left frontal cortex activates in positive emotions, and the right frontal cortex activates in negative emotions. As we already know, the prefrontal cortex plays the central role in actor-centered decision-making and in actor-centered appraisal of situations. The function of the prefrontal cortex is to calculate "what is good for the organism," more than to calculate "what is true" in an abstract, dispassionate sense. To me this suggests that the emotional "affiliations" of the left and right cerebral cortex are secondary to the cognitive functions of the two frontal lobes.

If this is so, then the brain regulates emotions through a

simultaneous integration of "vertical" and "horizontal" circuits. The two amygdalae are in charge of the instantaneous (and to a large extent hardwired or prewired) emotional response and the two frontal lobes are in charge of emotional reactions based on rational, cognitive analysis. These two inputs into our emotional responses, one rational cortical, the other instinctive subcortical, are combined in the fronto-amygdaloid circuits, this producing vertical integration of emotions. At the same time, the interaction between the "positive" left and the "negative" right fronto-amygdaloid circuits via the corpus callosum and the anterior commissures produces horizontal integration of emotions.

Driven to Discovery

So far in this chapter we have discussed the relationship between emotion and cognition, and how both are tied to the functional differences between the two hemispheres, in a somewhat abstract way. But clearly, different people have different emotional and cognitive styles. It is time now to examine the relationship between individual cognitive styles and individual emotional styles, and how they relate to the two cerebral hemispheres.

To begin considering this relationship, let's try to imagine a Prozac-popping Ferdinand Magellan or a Prozac-popping Christopher Columbus. It has been said that had Prozac been available in the days of the great mariners, they would have popped a pill or two and stayed happily partying in Seville, Lisbon, or Cádiz instead of embarking on their momentous explorations. Thankfully, these images are merely figments of my imagination because had it corresponded to any historical reality, America may have never been discovered by Europeans or the international dateline never introduced.

The image of a Prozac-popping Columbus is fanciful but it captures an important truth: Any quest for radical innovation, any voyage into the unknown, is driven by a feeling of dissatisfaction with the status quo. It is a restless, pressured feeling, consistent in affective tone with the one ascribed to the right hemisphere. The quest for exploration, for novelty, for *what ought to be*, goes hand in hand with the brooding dissatisfaction with *what is.* Perfectly content people do not discover new lands, do not circumnavigate the globe, and do not create revolutions in science. If everything is hunky-dory, why bother?

The folkloric image of a pioneer is not one of a happy type but of a brooding type. Bipolar disorder and bouts of depression are known to be the fate of many a great writer, scientist, and explorer. Psychologist Kay Redfield Jamison, herself afflicted with manic-depressive disorder, writes poignantly about the connection between creativity and psychiatric illness.

Jablow Hershman and Julian Leib have termed manic-depressive disorder "the key to genius." In a series of fascinating books, they explore the role of manic-depressive disorder in the creative lives of both great heroes and great villains in history. They contend that great contributors to human civilization, such as Beethoven, Byron, Dickens, Newton, Pushkin, Schumann, and van Gogh all suffered from this disorder. Churchill also suffered from bouts of depression, his infamous "black dogs," and the ferocious pace of literary production in his life is suggestive of hypomania. (Michelangelo was known to suffer from depression, but it is not known whether he also had bouts of hypomania.)

But manic-depressive disorder also played a role in the dark genius of aggressive and expansionist political and military empire builders. According to Hershman and Lieb, Napoleon, Hitler, and Stalin were all afflicted with this condition from the earliest years of their exploits. So was the Russian prince Potemkin, Catherine the Great's favorite and de facto prime minister, known for his infamous Hollywood-set-style Potemkin villages but also for his exceptional industriousness and effectiveness.

Both some of these heroes (Newton, Churchill) and villains

(Stalin, Hitler) appeared in the earlier chapter concerned with cognitive decline in historical personalities. The presence of affective disorder and of cognitive decline in the same individuals may be more than coincidental. Extensive scientific evidence exists that a lifelong history of depression is also a risk factor for dementia.

Accounts linking creativity to manic-depressive traits are abundant but anecdotal. I am not aware of any rigorous statistics relating mild affective disorder to acknowledged genius. To assemble such statistics, one would need to consider all the instances of genius afflicted with affective disorder and all the instances of genius free of this affliction, and compute a ratio between the two values. One would then compare the ratio to a similarly computed ratio in the general population. This is a task of daunting proportions for a number of reasons, not the least among them deciding whom to count as a genius (unless one relies on a survey akin to Charles Murray's ranking of the world's foremost historic personalities), and it may never be undertaken.

But two researchers from Stanford University, Connie Strong and Terence Ketter, have come very close to proving the point through less extravagant means. Using various psychological questionnaires, they found healthy people endowed with artistic creativity to be much closer in their personality makeup to patients with mild manic-depressive disorder than to ordinary healthy people. The researchers concluded that "negative-affective traits," which include mild nonclinical forms of depression and bipolar disorder, are strongly correlated with the capacity for creative ferment. In a similar vein, it has been shown that novelty-seeking is a particularly common trait among people with bipolar disorder.

This, of course, raises interesting questions: What are the brain mechanisms of manic-depressive disorder? What is its functional neuroanatomy? If you recall, functional neuroimaging and lesion data suggested a link between both left-hemispheric dysfunction and depression and right-hemispheric dysfunction and mania. What about manic-depressive bipolar disorder, which is different from both unipolar depression and unipolar mania? Preliminary evidence exists that brain-activation patterns can be dramatically different in the same patients during manic and depressed states. But these are extreme states. What about the stable traits of the bipolar patients' brain activity, the prevailing activation profile present in patients suffering from manic-depressive disorder most of the time? Available evidence suggests that the prevailing brain-activation profile of these patients is akin to the profile seen in depression and differs from the one seen in mania. It is dominated by an under-activation of the left hemisphere, while the right hemisphere shows normal activation patterns.

This physiological profile corresponds to the psychological state of persistent, nagging dissatisfaction with the way things are, producing the penchant for changing things. Periodic hypomanic states energize this penchant beyond the level seen in most people. The combined effect of background dissatisfaction and intermittent surges of energy is what fuels and drives creative accomplishment.[3]

But a personal emotional tone is not necessarily a constant. A case can be made that it changes through the life span and that true "emotional seasons of the mind" exist. Being at peace with oneself is, no doubt, a highly desirable state of mind, but in a young person it can be disappointing; it reeks of undue complacency, of premature aging, of unfulfilled callings, of mediocrity, even of outright anomaly. The romantic image of youth implies a measure of discontent and a measure of disquiet, the inner

---

[3]Several types of bipolar disorder exist. Bipolar I disorder is the most severe type, having severe manic episodes interspersed with severe depression, often accompanied by psychosis. Bipolar II disorder is less severe, with hypomanic episodes instead of full-blown mania. Cyclothymic disorder is the least severe form, with only mild hypomanic and depressive episodes, which are subclinical rather than clinical.

tension that fuels daring and rebellion. A cursory survey of the past few decades' political cataclysms shows that students were at the heart of many (if not most) of them—from the mass demonstrations in the United States and France in the 1960s to China's Tiananmen Square in the '80s and Indonesia in the '90s. This is a dramatic expression of the affective tone of the right hemisphere.

But as we age, the optimal affective tone changes. Research shows that with aging and with the movement well into the second half of life, negative emotions are emphasized less and the dominant affective tone becomes more positive. This is reflected in our brain activity: With age the amygdala becomes less active in response to emotionally negative stimuli while its response to emotionally positive stimuli remains unchanged. Thus, the balance shifts in favor of positive emotions, and as we age the affective tone of the left hemisphere becomes the norm. Our intuitive cultural perceptions agree with these findings. A restless octogenarian is often perceived, fairly or not, as the epitome of life unfulfilled, of a life cycle not completed, of striving for "too little too late." Being at peace with oneself in old age is the cultural stereotype to which most of us aspire.

This may sound counterintuitive, since depression is among the known ravages of old age. True, the prevalence of depression increases with aging, but so does the prevalence of osteoporosis, cancer, immunosupression, hair loss, and any number of other physical ailments. The association between aging and depression is not a specific one. It is but one of many manifestations of the finite nature of life and of increased susceptibility to any number of illnesses as life runs its course. Being at peace with oneself is the attribute of normal aging. Geriatric depression is not.

So the progression of the seasons of our mind, from the leading role of the right hemisphere in youth to the leading role of the left hemisphere as we age, unfolds in concert across multiple fronts: both where our cognition and where our affect are concerned. The right-to-left shifts of the center of cognitive gravity and of emotional gravity unfold hand in hand. This is a compelling expression of the unity between cognition and emotion in our mental life and our mental development.

13

THE DOG DAYS OF SUMMER

Cartographers of the Brain

What are the effects of aging on the two halves of the brain? Are they same? In the extensive and growing literature on the neurobiology of aging, these questions are hardly ever asked. And not surprisingly so, since nothing in our traditional understanding of brain function suggested that any such differences might be expected. But once we recognize the lifelong right-to-left shift of cognitive gravity, the question that until now sounded a bit outlandish now finds its justification. The shift implies that the left hemisphere is being used increasingly throughout one's life span and that this trend extends well into advanced age. Does this disparity of use affect the rates at which the two hemispheres age? And if so, how?

These questions bugged me enough to make me forgo my creature comforts on a particularly rainy and miserable morning in June of 2003. The *New York Times* had branded this summer the rainiest in a hundred years. There is an expression in my native Russian language to describe it, "dog weather," but that day the weather was too foul even for a dog. The downpour was so heavy that my bullmastiff Brit refused to take our usual walk. The moment we left the building, he squatted on the pavement, turned around, and pulled me back ferociously into the dry indoor comfort. Not surprisingly, humans enjoyed the misery outside even less, and I had a string of patient cancellations.

Liberated from my multispecies responsibilities, I decided to turn adversity into an opportunity. I found the largest umbrella I could and walked to the Marriott Marquis hotel, where the Human Brain Mapping: 2003 Conference was taking place.

In technical parlance, brain mapping is known as *neuroimaging.* The term refers to a whole collection of techniques allowing scientists to measure and actually visualize various aspects of the brain through direct scanning. Some of these methods I've already addressed. The physical principles underlying these techniques differ, but they all provide information about either brain structure or brain physiology. The difference between the two is like the difference between still photography and a movie. Structural neuroimaging provides snapshots of brain structure. By contrast, functional neuroimaging offers a glimpse into the brain in action. Computerized Axial Tomography (CT or CAT) and Magnetic Resonance Imaging (MRI) are examples of structural neuroimaging informing us about brain morphology. Functional Magnetic Resonance Imaging (fMRI), Positron Emission Tomography (PET), Single Proton Emission Computerized Tomography (SPECT), and magnetoencephalography (MEG) are examples of functional neuroimaging informing us about brain activity. Some of these techniques, like CT scan, MRI, PET, and SPECT, have become relatively commonplace tools of clinical practice. Others, like fMRI and MEG, still mostly remain the exclusive territory of neuroscience research.

The advent of neuroimaging truly revolutionized mind–brain science and helped graduate it from its tenuous position among "soft" science wannabes to the ranks of mature, recognized sciences. As I've already stated, the impact of neuroimaging on cognitive neuroscience has been likened to the impact of the telescope on astronomy. If the brain is the "microcosm" of human cognition, emotion, and consciousness, then the telescope analogy is both apt and revealing. My own work with functional neuroimaging is conducted mostly through collaboration with various colleagues, and I try to follow the developments in this new field as closely as I can. The Human Brain Mapping: 2003 Conference was a terrific opportunity to get an aerial view of where the field was going.

The brain mapping convention floor at the hotel was brimming with energy. I was struck by the number of young scientists from all over the world and by their diversity. Over the years of attending scientific conferences of various types, I have noticed that different disciplines related to mind–brain sciences—psychology, psychiatry, neuroscience, computer science, philosophy—attract different human types. It may be an illusion, but I don't think so. (Whether different personalities gravitate toward different disciplines, or whether different disciplines shape different personalities, or both, could be the subject of an interesting sociological project.) At the conference I was struck that they were all there, a truly multidisciplinary meeting of the minds, so to speak. As well it should be, since cognitive neuroscience today finds itself on the overlap of all of these disciplines.

Most scientific meetings have two kinds of presentations: talks and posters. Somehow, talks are considered more prestigious. They are certainly less laborious: You get on the podium, say what you have to say, and you are done. With posters, you have to prepare a visual display of your work, attach it to a board, and stand in front of it, usually for a few hours, explaining its content to the wandering spectators parading back and forth through the crowded aisles of the convention hall. In presenting my own work, I have always preferred talks to posters. But for a consumer of other people's work, posters offer a far more efficient way of taking in a lot of information fast. So instead of going to the talks, I paced through the endless aisles lined up with posters (almost two thousand of them in all), taking notes in my handheld computer.

Not surprisingly, my primary interest was new information on what happens to the brain as it ages. It is well known that the sulci and ventricles become larger with age, suggesting brain atrophy ("shrinkage" in the popular vernacular), with the frontal lobes particularly vulnerable. But these are very general observations. Is there anything more specific? I have a pretty well-articulated hunch about the fate of the two hemispheres in aging, but it requires confirmation—or refutation. This is one of the reasons I am here.

Sulci are among the most visually striking features of the brain, giving it its distinctive walnut appearance. Sulci are like deep canyons squeezed between mountain crests, the gyri. The evolution of the mammalian brain is remarkable for its progressive fissurization, an increasingly intricate and complex landscape of gyri and sulci, interwoven into an almost rhythmic arrangement. The evolutionary pressures behind this development probably relate to the growing surface area of the cortical mantle. As its size continued to increase throughout mammalian evolution, a "smooth" brain would have required an increasingly large cranial vault, and consequently an increasingly large skull. Imagine a creature with a whale-size head mounted on a human body. Smart maybe, but not very mobile and not very pretty. Instead of this self-defeating blueprint, evolution came up with a method of packing a very large cortical surface area inside a head kept on a human scale, so to speak. The method was to let go of a smooth cortical surface and instead to crumple and crease it like a walnut. An ingenious device worthy of a patent, but arrived at through the teleologically blind forces of mutation and natural selection.

But what happens to the walnut as we age? This was addressed in a poster showing the decrease of sulcar depth in normal aging. The study, conducted at Johns Hopkins University and the National Institute on Aging, uses an elegant methodology originally developed for earth studies and applied here to the studies of the brain. Interdisciplinary crossfertilization at its best—and at its most far-flung! It turns out that the sulci, the spaces between the cortical convolutions, become shallower as we age. This implies the atrophy of surrounding cortical tissue. Imagine a canyon becoming shallower and shallower over time, due to a gradual erosion of the surrounding cliffs. An elegant study yielding an expected finding, but this is not the end of it. The study has shown that the shallowing of the sulci is far from uniform. It is particularly pronounced in the parietal and occipital regions of the right hemisphere. By contrast, the sulci in the left hemisphere exhibit less change with age.

I moved to another MRI study of the aging brain. The authors of the study are from Australia, one of my favorite places on earth, so I was delighted to see their work represented. The study focused on the *insula*, a phylogenetically old region ("paleocortical" rather than "neocortical") hidden deep in the bottom of a canyon formed by the frontal, temporal, and parietal lobes. The function of the insula remains somewhat of a mystery, so much so that in many neuroanatomy and neuroscience texts any mention of the insula is conspicuous mostly by its absence. Traditionally, the insula has been implicated in putting together the senses of smell and taste. But its strategic location and rich connectivity suggest that it is involved in much more, perhaps in integrating the information about the organism's body and its internal states, and possibly interfacing this with information about the outside world. The Australian poster showed that the amount of gray matter in the insula of the left hemisphere does not particularly change over time, but in the right hemisphere it declines with age considerably.

The next poster to catch my attention was from Japan. It was an MRI study of aging male brains, using *voxel morphometry*. The "voxel" is to neuroimaging what the "pixel" is to the television screen—the smallest spatial unit of analysis. The existence of such a unit allows all kinds of quantitative analyses of brain-imaging data. You can, for instance, count the number of voxels contained within the image of a particular brain structure and by so doing express its size as a number. This is precisely what the Japanese scientists did, comparing the sizes of various brain structures across the fourth, fifth, and sixth decades of life. Decline in size with age is evident in several neural structures, and it starts earlier in the right hemisphere than in the left hemisphere. In the right hemisphere, the gray matter decay becomes evident already during the fourth decade of life and affects a number of structures during the fifth decade. In the left hemisphere, the decay barely begins during the fifth decade.

Another poster dealt with aging and depression. The MRI measurements of elderly individuals with depression show a size reduction of the frontal lobes and of the right (but not left) hippocampus, a structure closely linked to memory, even though not a site where actual memories are stored.

I moved on and saw a poster with an intriguing title about age, gender, handedness, and brain volume. The study was conducted at UCLA, at one of the best-known neuroimaging centers in the world. It focuses on young people; there is some gray matter loss already between the relatively tender ages of eighteen and thirty. The authors do not comment on any differences between the hemispheres. But as I examined the numbers on the poster, it looked to me like the decline is greater, ever so slightly, in the right hemisphere than in the left hemisphere.

The Steely Left Flank

Five studies presented at the meeting showed a greater atrophy in the right hemisphere than in left hemisphere with age, and not a single one showed the opposite pattern. How can we explain this disparity in decline? We already know that with age the left hemisphere takes an increasingly important role in our mental life, while the contribution of the right hemisphere continues to recede. Could it be that the disparity in use causes the disparity in decay, where more use equals less decay? (The relationship between neural use and neural protection will be sorted out in the next chapter.)

Before we are ready to draw conclusions, let us show some restraint and do a little math. Let us consider the so-called binomial distribution, the same distribution that you would use to compute the probabilities of certain outcomes in flipping a coin. I went to the convention with a precise hypothesis: that age-related brain atrophy is asymmetric, affecting the right hemisphere more than the left hemisphere. I made it my business to attend every poster session and survey all the posters presented at the meeting. I can feel confident that my sampling of the posters did not reflect any particular bias. If we discard the studies unconnected with atrophy failing to compare it on the two sides of the brain or showing equal amount of atrophy, we are left with exactly five studies showing asymmetric atrophy. We have just reviewed these studies.

The chance probability that the atrophy shown in the first of these studies was greater on the right side than on the left is 0.5. And the chance probability of this outcome for every subsequent study is also 0.5. So the probability that all five studies would show more atrophy in the right hemisphere than in the left, and not a single study showed an opposite outcome, merely by chance is 0.5 to the power of 5. This equals 0.0313, an exceedingly small number, barely more than three in a hundred. (For the mathematically pedantic among my readers, let me point out that this calculation reflects the fact that I had a very specific a priori hypothesis: greater atrophy in the right hemisphere than in the left hemisphere. Without such a specific hypothesis, the probability of all five studies showing the same direction of asymmetry, either left or right, by mere chance would have been 0.5 to the power of 4, which equals 0.0625, still a pretty small number.)

The most commonly accepted statistical convention used by scientists dictates that any event with the likelihood of chance occurrence below 0.05 (five in a hundred) is assumed not to be a chance event, but rather to reflect a genuine regularity. So by this statistical "gold standard" my "catch" on the brain mapping poster floor is extremely unlikely to be a chance fluke. Quite the contrary, it is extremely likely to reflect a genuine brain phenomenon.

The notion that the right hemisphere ages faster than the left hemisphere has been around for some time, but there were no solid data to support it. The assumption rested mostly on the changes in neuropsychological test performance with aging. But the neuroanatomical interpretation of these changes was tenuous at best. It has been known, for instance, that on the Wechsler Adult Intelligence Scale (WAIS), the Performance IQ (PIQ) declines more rapidly with age than does the Verbal IQ (VIQ). But the commonly made assumption that Verbal IQ reflects the functional capacity of the left hemisphere and Performance IQ reflects the functional capacity of the right hemisphere is patently wrong for a number of reasons.[1]

So, despite the numerous claims to the contrary, different rates of VIQ and PIQ decline with age tell us very little about the fates of the two hemispheres at later stages of life. Other attempts to use neuropsychological observations to chart the rates of decline of the two hemispheres in aging have been equally arbitrary or flawed. But now, finally, we have direct neuroimaging evidence to show that the right hemisphere indeed declines more rapidly with aging than the left hemisphere. Or, to put a more positive spin on the data, we have direct evidence that the left hemisphere withstands the effect of age-related decay better than the right hemisphere.

---

[1]Some of these reasons include the following: (1) The WAIS subtests used to measure the Performance IQ are timed, while the subtests used to measure the Verbal IQ are not timed. Since speed of mental and physical operations often declines with aging, the two groups of subtests are affected differently. (2) Most of the Verbal IQ subtests greatly depend on education and cultural background, while most of the Performance IQ subtests are relatively culture free and education free. Therefore the person's cultural and educational background will unequally affect performance on these subtests in a way that has nothing to do with the physical integrity of his or her brain. And so on.

14

USE YOUR BRAIN AND GET MORE OF IT

New Evidence for New Neurons

Why is it that the right hemisphere decays more rapidly than the left hemisphere in aging, and what protects the left hemisphere from decay, making it an "evergreen" of sorts in the seasons of the mind? What is the biological basis for that mysterious disparity between the two halves of the brain? Is it possible that as the brain ages it also renews itself and that this process of renewal is, for some reason, more vigorous in the left hemisphere than in the right hemisphere? To help answer these questions, I returned to the convention floor and kept walking up and down the aisles looking for the data to test my other hunch.

"Use it or lose it" is a well-worn adage traditionally finding its meaning in the world of athletics. But lately it has found a new meaning in brain science. In the course of the last decade, spectacular discoveries have been made that changed our basic assumptions about what happens to the brain in the course of a lifetime and upended some of the most sacrosanct beliefs in neuroscience. As recently as two decades ago, we used to think that a human being was born with a fixed collection of nerve cells in the brain (neurons), which gradually died out as we aged without any possibility of regeneration. As a graduate student at the University of Moscow in Russia many years ago, I referred to this assumption (which was ideologically agnostic and prevalent on both sides of the Iron Curtain), jokingly and skeptically, as the NNN principle—"No New Neurons!"

Neuroscientists recognized that the NNN principle set the brain apart from the rest of human body, since most other organs have the capacity for regeneration. Neuroscientists also recognized that the NNN principle was not ubiquitous, since it has been known for years that the brains of several bird and rat species do have the capacity for regeneration.

For years a handful of iconoclastic scientists like Fernando Nottebohm and Joseph Altman were trying to draw the attention of the neurobiological community to these animal research findings and to their implications for human therapies. But their efforts were dismissed as irrelevant to the human brain. It was thought that humans were different, that the inability to regenerate new neurons was the price that we had to pay for the privilege of hanging on to the old neurons, the neurons that encoded our previously acquired knowledge, our memories, our selves.

On the surface, this sounded like a plausible exercise in "neuroteleology," since, as we have abundantly established, humans depend on previously accumulated or learned knowledge far more than any other species. But on closer scrutiny the argument does not hold up, since we lose our old neurons anyway in the course of life, whether we like it or not. Neurologists and neuropsychologists know very well that even in perfectly healthy people CT or MRI scans of the brain look differently at different ages, suggesting some degree of neuronal loss. As we already know, in normal aging the neuronal loss seems to occur both in the neocortex, where the pattern-recognition generic memories are contained, and in certain subcortical structures and around the ventricles, the cerebrospinal fluid–containing cavities deep inside the brain. Since the neocortex is clearly not entirely spared, the only explanation of how we endure neocortical neuronal loss without the loss of essential, previously accumulated knowledge is by assuming that our memories, particularly the generic memories, are stored in a highly redundant fashion. Such redundancy is reflected in, among other things, the "pattern expansion" discussed in the previous chapters.

The NNN axiom, regarded as ironclad for decades, finally became indefensible with the work of Elizabeth Gould and others, who have demonstrated the existence of ongoing neuronal proliferation in several monkey species. Monkeys are too close to humans to dismiss such findings as irrelevant, and the monkey findings are particularly exciting because they show the proliferation of new neurons in the heteromodal association cortex of the frontal, temporal, and parietal lobes. It was also shown that new neurons continue to grow throughout the life span in the hippocampi. All these parts of the brain are especially important in complex cognition, and they are particularly vulnerable both in normal aging and in various forms of dementia, including Alzheimer's disease. Potentially, the findings of lifelong neuronal proliferation in the neocortex and in other parts of the brain (including the hippocampi, so important in the formation of new memories) open the door for a wide range of therapies in humans.

Today, we know that the old premise of "No New Neurons!" is simply not true. New neurons constantly develop out of stem cells throughout a lifetime, even as we age. So our brain has the ability to restore and rejuvenate itself. Contrary to long-held beliefs, neurons do not stop developing in infancy. Far from it; they continue to grow throughout the whole life span, well into adulthood and even into advanced age.

Furthermore—and this is particularly important—there has been growing evidence that the rate of development of new neurons could be influenced by cognitive activities in a way not dissimilar from the manner in which muscle growth can be influenced by physical exercise. This was demonstrated with particular clarity in experiments conducted at the Salk Institute, one of the premier centers of biomedical research in the world. A much greater rate of new neuronal development (up to 15 percent more) was noted in mice immersed in an environment filled with toys, wheels, tunnels, and other "mouse-brain teasers," than in idle mice left to their own devices. The mice from the enriched environment have also shown significant advantages on various tests of rodent intelligence. The neuronal proliferation triggered by cognitive exercise was especially pronounced in the hippocampus. The finding is of paramount importance because, as we have seen, the hippocampus is particularly important in memory and is among the brain structures most affected at the very early stages of Alzheimer's disease. Not surprisingly, the levels of chemicals stimulating the growth of new neurons in the brain also increase as the result of exercise. This was demonstrated for the Brain-Derived Neurotrophic Factor, or BDNF for short.

While much of the early evidence was obtained in animals, direct human evidence is also beginning to appear, causing great excitement in the scientific and biomedical communities.

Some of the recent findings are truly dramatic. It has been shown, for instance, that new neurons continue to appear in the adult human hippocampi. This finding, first reported by the Swedish scientist Peter Eriksson, has become frequently quoted in neuroscientific literature. What's more, new neurons continue to proliferate not only in healthy brains but also in the brains of patients suffering from Alzheimer's disease. Findings like these certainly breathe new life into the "use it or lose it" adage. One is tempted to rephrase it, "Use it and get more of it."

The notion that mental activities can actually change the brain is gaining an increasing number of supporters in the scientific and biomedical communities. Much of the recent work on the subject has been reviewed in an excellent book by Jeffrey Schwartz and Sharon Begley, *The Mind and the Brain*. But what exactly happens in the human brain as a result of vigorous mental activity? If you asked me this question a decade ago, I would have said that the connections between the neurons become more numerous and stronger. This would imply a more vigorous growth of dendrites and synapses, and the development of extra receptor sites, to which the neurotransmitter molecules bind. I would have also said that the small vessels carrying blood (and through it oxygen) to different parts of the brain proliferate.

I still say all of these things. But the past decade brought new, even more stunning discoveries about the brain's plasticity and how it continues to be molded by environment throughout the lifetime. We know this from animal research, which brought about a true revolution in our thinking about the life of the brain. As we have already learned, cognitive exertion increases the rate with which new neurons appear in a wide range of brain structures, which may include the prefrontal cortex, a brain region particularly important for complex decision-making, and the hippocampi, the sea horse–like structures particularly important for memory.

Since all mammalian brains operate on the same fundamental neurobiological principles, we could reasonably assume that the human brain is also capable of producing new neurons throughout the life span. But is there direct evidence of this happening, and can the rate of new neuron production be increased by cognitive exercise in humans as well? This proposition would have sounded so outlandish even a decade ago, and certainly two decades ago, that I probably would have felt my own intelligence insulted by a mere consideration of this possibility. And I would have been wrong!

The first evidence that brain structures may actually grow, actually increase in size as a result of environmental factors even on the macroscopic scale, came from none other than . . . cab drivers. The finding is especially striking because of its simplicity and direct explanatory relevance. Hippocampi were found to be especially large, larger than in most people, in London cab drivers, whose job requires the memorization of numerous complex routes and locations. Since the hippocampi are so important in memory, and good cab drivers in a huge city like London must memorize a particularly large number of spatial routes and locations, they strain their hippocampi, so to speak, more than most people, just like a weightlifter strains his muscles more than most people. Furthermore, the longer the cab drivers were on the job, the larger were their hippocampi: The size of the hippocampi was directly proportionate to the number of years on the job. This suggests a direct relationship between the amount of a certain type of cognitive activity and the size of a neural structure involved in this activity.[1]

The cab-driver findings are remarkable in several respects. First, an important neural structure can continue to grow well into adulthood. What's more—and this is particularly important—the growth of a neural structure appears to be stimulated by its use. More years on the job generally implies older age, which in turn would suggest hippocampal atrophy. Yet here we have older people with larger hippocampi due to increased mental activity of a particular kind. The effects of vigorous cognitive stimulation seem to offset and override the detrimental effects of aging—perhaps to a substantial degree.

While cognitive exercise stimulates the proliferation of new hippocampal neurons, other factors may retard it. As it turns out, neuronal proliferation in the adult hippocampi is a process both delicate and resilient. It can be upset by, among other things, brain inflammation, a condition found in diseases as

---

[1]As a carless Manhattanite constantly hailing taxicabs and finding myself directing cab drivers on the verge of getting lost, I would not be too optimistic about our ability to replicate the London findings in New York. But the Old World is a different place, where driving a cab, like waiting on restaurant tables, seems to be regarded and accepted as a profession in its own right, rather than a stopgap between a failed acting career and a winning lottery ticket.

diverse as Alzheimer's disease, Lewy body dementia, and AIDS Dementia Complex. (This is probably due to the disruptive effect of inflammation on the brain stem cells, the "prefab" cells that subsequently differentiate into a variety of specific neurons.) But adult neurogenesis in the hippocampi is restored when the inflammation is reduced.

Having established that cognitive exercise spurs the growth of new neurons, we are ready to ask our next question: How specific are these effects? The brain is a diverse, heterogeneous organ. Different parts of the brain are in charge of different mental functions, and different mental activities call upon different parts of the brain. If mental exercise, the use of one's brain, stimulates the growth of new neurons, then it is quite plausible that different forms of mental activity will stimulate such growth in different parts of the brain.

For instance, is hippocampal enlargement specific to those activities that must rely on spatial memory, or is it the case that certain brain structures are sensitive to the effects of any mental stimulation and other brain structures are not? What would be the effects on the brain of other types of mental activities, which rely on vastly different cognitive functions? To expand on the thought, if the hippocampi are enlarged in cab drivers, can we reasonably expect that the left temporal lobe (the language lobe) would be enlarged in a writer, the parietal lobes (the spatial lobes) in an architect, and the frontal lobes (the executive lobes) in a successful entrepreneur? Or is it the case that certain structures, possibly the hippocampi among them, will be enlarged in any profession requiring mental exertion regardless of the specifics, and certain other structures will not be?

Since different types of cognitive exertion call into action different parts of the brain, it would stand to reason that they will also stimulate extra neuronal proliferation in different parts of the brain. Therefore, the idea that the brain-stimulating effects of cognitive activities are at least somewhat specific is not totally outlandish. In fact, the more one thinks about it, the more plausible it sounds. But plausible or not, do we have direct evidence to this effect?

The Bilingual Brain—and the Musician's Mind

Spectacular as the London cab driver finding was, for a while it remained one of a kind. And one study was not enough, precisely because of the finding's spectacular nature. The more ambitious a scientific claim is, the more profound its implications, the higher the bar is set for its acceptance by the scientific community, and the more rigorous proof that is required. This is one of the most inviolate rules of science, and the cab-driver findings were received with a degree of caution.

So you can imagine my excitement when in the course of a few hours on the brain mapping conference floor I stumbled into not one, but two similar findings, both involving MRI. In the spirit of the meeting, they came from two very different corners of the world.

The first study, conducted at the Wellcome Department of Imaging Neuroscience of the Institute of Neurology in London, involved MRI measurements of the size of the *angular gyrus*, a cortical area where the temporal, parietal, and occipital lobes come together. It is part of the heteromodal association cortex, in charge of integrating inputs arriving from multiple sensory channels: visual, auditory, and tactile. The angular gyrus of the left hemisphere plays an exceptionally important role in language, particularly in processing various relational constructs: before/after, above/below, left/right, passive voice, possessive case, and so on. We know all this because of extensive observations of what happens when the left angular gyrus is affected by a brain lesion, such as after a stroke, or by a gunshot wound. Damage to this part of the brain produces severe language impairment, a form of aphasia of a particular kind. The angular gyrus is among most researched parts of the brain, and its functions have been described in numerous scientific articles and books, including the classic monograph by my mentor Aleksandr Luria, *Traumatic Aphasia*.

The author of the Wellcome study, a young man pacing a bit nervously in front of his poster, offered to explain it, and within seconds we were engaged in an animated discussion. It turns out that the left angular gyrus contains significantly more gray matter in bilinguals (people fluent in two languages) than in monolinguals (those fluent in only one language). Furthermore, the white matter underlying it is characterized by greater density. In plain English this means that there are more neurons and more connections in the left hemisphere of the individuals in command of two languages than in the people who speak only one language.

Being a bilingual (trilingual in fact, but let's not push it), I congratulated myself on possessing a large left angular gyrus and began to ponder the significance of the study. Gray matter consists of neurons and the short, local connections between them. The findings suggest that extra cognitive activity triggers an increase in the number of neurons in the cortical regions doing the work. It also suggests that extra cognitive activity stimulates the growth of short, local connections between neurons.

Neurons are not born right where they perform their function. They are manufactured around the walls of the lateral ventricles as undifferentiated stem cells. Then the stem cells differentiate into specific types of nerve cells and migrate to their ultimate destinations in various parts of the brain, including the neocortex, far away (in terms of brain space) from their birthplace. So it appears that neuronal migration traffic is regulated, at least to some extent, by cognitive activity, which determines not only how many new neurons should be manufactured but also where they should go.

But this is not all. Not only do bilinguals have more gray matter in their left angular gyrus than monolinguals, but they also have greater white matter density in the left hemisphere. White matter consists of long myelinated pathways in charge of connecting far-flung cortical regions. It appears that extra cognitive activity stimulates the growth of long-distance pathways as well. This is no less important than the number of neurons, since the complex functions of the brain arise from multiple interactions between huge numbers of neurons, both nearby and far removed from one another, and such interactions are mediated by the pathways between neurons. The denser the matrix of such pathways, the greater the functional capacity of the neuronal network. What's more, bilinguals appear to have greater white matter density than monolinguals not only in the left hemisphere but in the right hemisphere as well. This finding suggests that the right hemisphere plays a role in learning a second language, which resonates with the functional neuroimaging studies of bilingualism discussed earlier in the book.

The study is truly a gem, not least because it involved both early bilinguals (who acquired the second language early in life) and late bilinguals (who acquired the second language later in life). An increase in left-hemisphere gray matter compared to monolinguals was evident in both groups of bilinguals. This means that the brain-enhancing effects of cognitive activity are not limited to young age. They continue much later in life as well.

The next study compares the size of a cortical area known as the *Heschl's gyrus* in professional musicians and nonmusicians (many of us will fall into the latter category). This cortical area is critical for sound processing. And guess what—the Heschl's gyrus is twice as large in musicians than in nonmusicians. Furthermore, the greater the intensity of practicing music in the last ten years, the greater the size of the Heschl's gyrus. Again, the relationship between cognitive activation and specific brain regions is apparent and striking.

And then, a few months later, an MRI study of brain changes in jugglers was reported in *Nature* magazine, one of the most respected science journals in the world. Healthy volunteers, none with prior experience in juggling, were trained for three months in a three-ball juggling routine. As a result of the training, the volunteers achieved enough juggling proficiency to keep the balls in the air for at least sixty seconds. When their before and after brain MRI scans were compared, it turned out that the amount of gray matter increased in the temporal lobes in both hemispheres and in the parietal lobe of the left hemisphere. With the interruption of practice, the effect gradually dwindled away and the gray matter gains in the parietal and temporal lobes were reduced. This was evident in the third MRI scan recorded three months after the discontinuation of juggling practice. So the effects of skill practice on neuronal proliferation in very specific parts of the brain could be demonstrated even within a relatively short period of time.

A devil's advocate might say that musicians become musicians because they are *born* with a larger Heschl's gyrus, which in turn endows them with a particular musical talent. And couldn't it be that a natural selection occurs among the cab drivers, such that those born with larger hippocampi find the job more agreeable because they have a better memory for complex routes? And couldn't it also be that people born with a larger left angular gyrus have a greater natural aptitude for languages and thus learn more of them? But while biology is a major part of our destinies, the "destiny imperative" does not explain everything. It cannot explain, for instance, why the sizes of the hippocampi, the Heschl's gyrus, and other parts of the brain are positively correlated with the amount of time spent in practicing certain cognitive skills. And it certainly cannot explain the rapid, and reversible, effect of juggling practice on the brain. These correlations indicate that ample room exists for pushing biology around, that biology sets a range of expressions (and not a fixed constant) for each ability, and that exactly where within this range we end up depends on us—on what we do with our brains and with ourselves.

The Aging Hemispheres and Dementia

So, by engaging in vigorous mental activities, we change our brain in ways so profound that certain brain regions may actually grow in size. The next issue to address is: which areas?

The relationship between the nature of mental activities of cab drivers, bilingual people, musicians, and jugglers, and the brain structures affected by the activities described in these studies, appears to be impressively specific. In order to remove any doubt in the specificity of the cognitive-stimulation effects on the brain, neuroimaging studies would need additional rigorous controls. By this I mean the measurements of certain additional brain structures with minimal or no involvement in the cognitive activities used for brain stimulation. And one would need to meticulously show that such control brain structures did not increase in size, that only the brain structures directly involved in the cognitive activities did. But the findings reviewed earlier in this chapter offer a good start.[2]

---

[2]Sometimes negative findings are as important as positive findings, particularly when the former help clarify the latter. It has been known for some time that neural stem cells line the area around the lateral ventricles of the brain. In rodents these prefabricate cells migrate from there to the olfactory bulbs found at the base of the frontal lobes (however meager they may be in a rodent), and this process continues throughout the animal's life span. But not so in humans. A group of American and Spanish scientists have shown that, like in rodents, in humans such prefabricate cells also continue to be born around the walls of the lateral ventricles even in adulthood. But unlike the rodents, in the human brain the prefabricate cells fail to migrate to the olfactory bulbs. "Immigration denied," concluded Yale's Pasko Rakic, one of the world's foremost neuroscientists and a skeptic when it comes to neural plasticity in the adult human brain.

But does this negative finding mean that stem cells don't migrate *anywhere* in the human brain? In claiming that humans differ from other mammals when it comes to adult neurogenesis, the skeptics invoke the argument that the human brain stands particularly to gain by conserving neural circuits rather than by modifying them, since we depend on previously accumulated knowledge more than other species. But while the latter is undeniably true, it

Let's now step back and think about the real-life implications of all these studies. Most of us, in fact all of us, exercise certain mental faculties more than others, by virtue of doing our jobs or enjoying our hobbies. This is a pervasive, universal fact of life. The effects of learning music or learning languages, or of learning complex street routes or juggling routines, on the brain are merely cases in point, examples of a profoundly general phenomenon. To the extent that the brain-stimulating effects of mental activities are even somewhat specific—and they appear to be—they are likely to benefit different brain structures in different people. But are there any invariants despite the differences? Are there any common themes dominating the brain-stimulating effects of mental activities, which rise above this sea of individual differences dictated by the diversity of our educations, occupations, and experiences?

Enter again the two cerebral hemispheres. We already know that most cognitive skills are controlled by the right hemisphere at the early stages of learning, but they are controlled by the left hemisphere once we reach a certain level of mastery. This means

---

is also true that we constantly acquire new information and relentlessly update our knowledge, and different parts of the brain contribute to these processes to unequal degrees. Suppose that the destination of stem cell migration from their birth area around the lateral ventricles depends on the level of neural activity in the target areas, the stem cells being somehow drawn to where the most action is. In that case, the general principle governing neural cell migration in adult brains will be expressed very differently in different species, since they rely on different brain structures for survival. There would be very little reason to expect that in humans the cells would end up in the olfactory bulbs, since humans are least dependent on olfaction, unless of course we are gourmet chefs or perfume designers. It would make much more sense to expect that the visual or auditory or complex association cortex would be the likely magnets for stem cell migration streams in the adult human brain. The difference between mammalian species may be less in the degree of stem cell migration and more in their neural targets. "Immigration regulated by workforce demands" may prove to be a more accurate way of capturing the stem cell migration processes in mammalian brains, including the adult human brain.

that with experience we increasingly rely on our left hemisphere across a very wide range of mental activities and skills, whatever these activities and skills may be in a given individual. It appears that, as we move through life, the brain structures housed in the left hemisphere become increasingly engaged compared to the brain structures housed in the right hemisphere. Therefore the left hemisphere becomes the predominant beneficiary of the enhancing effects of mental activities, regardless of their specific nature. (Of course, this conclusion is predicated on the assumption that the brain-stimulating effects of mental activity are at least somewhat specific, which at this point we have strong reasons to believe this to be precisely the case). With this in mind, it should come as no surprise that the practice-enhancing effects of activities as diverse as language and juggling were both seen particularly in the left hemisphere.

And so, as the brain mapping conference came to a close, I had a feeling (a bit complacent, but maybe not wholly undeserved) that I have glimpsed beyond the trees at an important corner of the forest. My take-home message from the meeting was, in fact, three messages wrapped in one:

- The right hemisphere decays more than the left hemisphere as we age.
- The left hemisphere benefits increasingly more than the right hemisphere from mental exercise, as we move through life.

Although not quite a formal Aristotelian syllogism, the following conclusion is warranted:

- The left hemisphere is better able to withstand the decaying effects of age because it continues to be enhanced and strengthened by cognitive activities as we age.

Earlier in the book we discussed the protective effect of education against dementia. With the knowledge available today, we may conclude with a fair degree of confidence that this is probably due to the fact that educated people are more likely to make their living with their brain than with their brawn, and thus will benefit more from the brain-enhancing effect of a lifetime of vigorous mental activity. And as we are approaching the end of this chapter, we may be tempted to conclude that such a protective effect will be more apparent in the left hemisphere than in the right hemisphere.[3]

Neuroscientists studying dementias have long been perplexed by the many faces of dementia at its early stages. The early manifestations of dementia, any dementia, are extremely diverse. This is especially the case with Alzheimer's disease. It is true that in the majority of patients the earliest manifestations of Alzheimer's dementia start with memory impairment, but in a significant number of such patients other functions suffer first: language, spatial orientation, or executive functions. Several neurologists, including one of the world's foremost experts on dementias, estimated that the earliest symptoms of cognitive decline involve memory in up to 70 percent of people eventually diagnosed with Alzheimer's-type dementia. But in at least 30 percent of such people (a huge minority) memory decline is preceded by the decline of other functions, such as language, spatial orientation, or executive functions, with the "personality change" suggestive of frontal-lobe disease.

When the diversity of the early symptoms of Alzheimer's disease was first recognized, hypotheses began to promulgate that Alzheimer's is not one disease but many separate diseases. This notion, popular in the 1980s, has since been discarded. It is more likely that the diversity of the early symptoms of dementia is the flip side of the diversity of the profiles of neural protection offered by a lifetime of certain kinds of mental activities. These

[3]You will recall that this is true for the right-handers among us and for the majority of the left-handers, but the reverse may be true for a minority of the left-handers. (See chapter ten for a reminder).

profiles obviously differ in individuals, depending on the nature of their lifelong activities. While some cognitive functions are exercised more (thus conferring neuroprotection on certain parts of the brain), other cognitive functions are exercised less (thus failing to confer neuroprotection on certain other parts of the brain). These latter brain structures will represent the "chinks in the armor" of neuroprotection, which will vary from person to person. Certain lifelong cognitive histories exercise particular parts of the brain more than others, and this may confer neuroprotection (albeit partial and temporary) on the exercised parts of the brain against the ravages of early dementia. This is only a hypothesis, but it is an intriguing one.

According to this logic, early dementia in a writer is less likely to affect language than spatial processes. In an architect, the disease progression will take an opposite course: language succumbing first and spatial processes much later. In an executive in charge of strategic planning, the frontal lobes will put up the longest resistance to the effects of brain decay. But in the proverbial London cab driver, memory will be the last to go, way after language or executive functions.

The brain structures benefiting from the neuroprotection conferred by exercise are able to withstand the assault of neurological decay for longer, maybe for much longer. A body of evidence now exists (and is growing) that aging individuals may remain functionally and cognitively sound despite the neuropathological signs of Alzheimer's disease and other dementias. Robert Katzman and his associates at Albert Einstein College of Medicine in New York and University of California, San Diego, studied a sample of such people and found that they had greater brain weight and more large neurons than the matched controls. It is likely that the unusually large brain weight was a reflection of a greater number of large neurons and pathways, which in turn was due to a lifelong history of cognitive vigor and exertion. This possibility, which even a decade ago would have been dismissed as fantastic, today finds support in observations such as those involving London cab drivers, bilinguals, and professional musicians.

A similar study, which I cited earlier, involved the much-researched nuns from the School Sisters of Notre Dame in Mankato, Minnesota. The nuns' lifestyle was remarkable in its mental richness and stimulation. The nuns were also remarkable for their longevity and mental vigor well into old age. It appeared as though Alzheimer's disease spared them. But when the brains of some of the nuns were examined after their deaths, the characteristic Alzheimer's tangles and plaques were found. The nuns were able to retain their mental powers despite the presence in their brains of the telltale neuropathological hallmarks of Alzheimer's disease. How was this possible? The most logical explanation is that the neuroprotection conferred by lifelong mental activities (extra neurons and the connections among them) was sufficient to counteract the effects of an otherwise dementing brain disorder and ensured the nuns' clarity of mind despite the biological markers of disease.

15

PATTERN BOOSTERS

Athletics, Art, and Einstein's Violin

Because of the recent scientific discoveries that mental activity can actually change the brain, I have become a firm believer in the merits and value of designing such activities in a systematic, rigorous way. I have been one of the early and vocal proponents of the idea that by engaging an aging individual in vigorous mental activities you may actually improve his or her brain's resistance to decay. Based on this idea, I developed a cognitive exercise program in New York City, which is alive and well and continues to attract participants as this book is being written. I have every reason to expect that our cognitive exercise program (to which we often refer as a "cognitive enhancement" or "cognitive fitness" program) will continue to be alive and well and to attract new participants in the future.

It has been known for some time that education seems to be a protective factor against mental decline and against dementia. This was an unexpected but in retrospect very sensible finding of the famous multicenter study on the keys to successful aging, the MacArthur Project. The assumption is that educated people spend a lifetime of more vigorous mental activities than less educated people, primarily because of the nature of how they make a living, and that such vigorous activity calls into action all the neuroprotective mechanisms discussed in the previous chapters. It does not take too great a leap of imagination to expect that a series of well-designed cognitive exercises based on sound neuropsychological rationale will be even more effective in stimulating neuroprotection than the workaday, often unavoidably haphazard, activities of a busy professional.

Whenever I introduce the concept of cognitive enhancement to the uninitiated public, I do so with conviction, but also with trepidation. At the heart of my timidity is less the issue of sufficient scientific evidence (I believe that it amply exists) than the concern about public skepticism. The notion of exercising one's mind may appear a bit far-fetched, not anchored in anything tried and familiar to a skeptical reader. Yet I will argue that mind exercise in its purest form is among the most ancient types of human activity and that we all engage in this activity much of our lives. For that, we need to turn to the mysterious function of art.

Two pastimes have been central to human civilization since the beginning of history: athletics and art. They have been central to virtually every culture and often go hand in hand. The ancient Minoans danced with their bulls (athletics and art intertwined) and portrayed them on the frescos of their elaborate labyrinthine temples in Knossos and elsewhere on the island of Crete. Ancient Egyptians left artfully decorated papyri with elaborate instructions on wrestling moves (again athletics and art intertwined). Ancient Greece, of course, provided the foundation for both the Western artistic idiom and organized athletic competition (the Olympic Games). And today seeing the latest Broadway show and going to the gym is equally de rigueur for a high-living Manhattanite.

Athletics and art are so organic to our culture that we take both of them for granted, without questioning their utility. In his book *The Mating Mind*—easily one of the most original, thought-provoking, and irreverent reads I have come across in recent years—Geoffrey Miller touches upon these questions from the standpoint of their evolutionary origins. But the evolutionary origins of a trait and its utility in modern society are not necessarily exactly the same; in fact, they most likely are not. So what *is* the utility of athletics and what *is* the utility of arts for us today?

The utility of athletics is intuitively clear. While not serving any specific practical purpose, physical exercises strengthen the body, instill discipline, and make us more prepared to face an open-ended set of possible physical challenges. Physical exercise also makes the vital systems necessary for survival—cardiovascular and pulmonary—more robust. So we intuitively and habitually embrace athletics as a useful ingredient of living, despite the absence of any specific practical purpose behind it. The origins of physical exercise may antedate the advent of our species. It can be argued that the precursors of athleticism are found in the rough-and-tumble play common among most mammalian species.[1] When my bullmastiff Brit has a predictable evening burst of activity, dashing about the apartment wildly urging me to join in, or when he tries to draw me into a friendly game of tug-of-war, I have to assume that these behaviors serve some adaptive role in the canine scheme of things. In his book Miller proposes that sporting competitions evolved as physical fitness self-advertising to the opposite sex. It has also been said that athletics channels male competitive impulses into nonlethal, nonviolent ritualized conflict. These may very well be parts of the evolutionary story of athletics, but hardly the full story of its utility today. For one, not all athletic pursuits are competitive. Believe me, when I prevail upon myself to climb into the swimming pool in my building's health club (only one flight of stairs

[1]A special role of the "pleasure hormones" endorphins released as a result of vigorous physical exercise and contributing to the feeling of well-being, such as "runner's high," is outside the scope of this discussion. Suffice it to say that the endorphin release following vigorous exercise may have evolved as a mechanism of positive reward for otherwise "useless" physical exertion.

away but a veritable test of my willpower), sexual self-advertising, or even channeling aggressive impulses, is the last thing on my mind, but the thought of a familial history of heart attacks and how to increase the odds that I will be spared one is the foremost.

But what about art? Art is equally ingrained in our lives, or even more so. Art permeates our lives so organically that we take it as a given, without concerning ourselves too much with its function or origins. Yet all attempts at understanding the origins and function of art in human civilization have met with rather limited success and resulted in less-than-compelling speculations.

It has been said that like science, art helps us understand the world around us. As a sweeping generality, this is probably true; but then again it is probably true for any human activity, making the claim a truism bordering on the platitudinous and thus explaining nothing. Like science, art is to a large extent a cerebral pursuit, or at least can be. But unlike science, art helps us understand the world only very indirectly. Just how indirectly becomes apparent once you realize that the veridical "true/false" distinctions cannot be applied to the creation of art the way they can be applied to claims made by scientists. And unlike science, art does not develop in a clearly discernible incremental progression. (Few people will dispute the assertion that twenty-first-century science is more advanced than nineteenth-century science, but can we say with equal conviction that contemporary art is more advanced than the art of the Renaissance, and that the latter is more advanced than the art of ancient Greece?)

Attempts have been made to find the origin of art in religious ritual. But even if this were so in the past (itself a difficult point to prove), the religious underpinnings of art would be hard to reconcile with the decidedly secular, even the arguably blasphemous poetry of Arthur Rimbaud and William Henley or novels of Salman Rushdie. Great art nonetheless! Furthermore, based on the premise of the religious origins of art, one would predict its decline in our increasingly secular times, but this is clearly not happening. One can even argue that religious precepts at times stifled art rather than promoted it. The prohibition of representational imagery in certain strains of Judaism, Christianity, and Islam is a case in point.

It has been proposed that unlike science, art conveys emotions and that therein lies the unique function of art. But drawings by M. C. Escher or etchings and lithographs by Yaacov Agam are hardly emotional; they seem to be quite intellectual, fruits of cerebrally constructed quasimathematical algorithms; seventeenth-century fugues are the embodiment of almost mathematical precision; and it is hard to get more analytical than Umberto Eco's prose or some early twentieth-century experimental poetry.

So elusive is the utility of art that it has even been proposed that its utility is in the very absence of intrinsic utility. A highly ingenious and provocative theory (but in my opinion less than entirely convincing) has been advanced that art, including music, is a surplus, "throwaway" activity whose sole function in society is to advertise its practitioner's mental fitness to prospective partners in the mating game. By extension, one could say that art is something that only highly successful societies can afford. Hence, the value of art is in the affirmation of the power of the society that has surplus resources to spend. By postulating the intrinsically useless nature of art, the "art for sex" theory (the moniker is mine) views art's only utility as a surrogate, as a marker of prowess in something else. Taken to its logical extreme, this position triggers the conclusion that art is even less than useless, that it may actually be harmful, "handicapping" by draining an inordinate amount of its practitioner's mental resources. This produces the paradox that art is a marker of ample mental resources precisely because the practitioner of art can afford to throw a lot of them away with impunity—just as the proverbial nouveau-riche showboat will parade his wealth with a hundred-dollar tip where two bucks would have sufficed. Ultimately, this position implies that much of our dazzling brain power, including the capacity for making and appreciating art, is to humans what a spectacularly colorful tail is to peacocks—nothing other than a resource drain to the point of being a burdensome, intrinsically useless amenity that evolved solely and exclusively as a net for sexual entrapment. The "art as sexual self-advertisement" theory, outlined by Miller in his book, offers an interesting perspective, but again the question of the difference between art's evolutionary roots, whatever they may have been, and its role in modern society comes up. The two may very well have diverged. Much as I enjoyed Miller's book, I think it is wrong to deny art any direct survival value to the species that invented it—to us humans. While provocative, this denial sounds to me like an extreme explanatory default bordering on explanatory desperation. In a similar vein, it has even been said that art is "biologically frivolous." But the only frivolous thing about this explanation of art is the explanation itself. Try harder!

One of the most perplexing things about art is that its forms are so numerous and so diverse that they defy any quest for a common denominator. What is the intrinsic commonality between Japanese calligraphy and a heavy metal concert? (What a bizarre thought—none, for God's sake!) Yet both are art, and we know art when we see it. A similar rhetorical question can be asked about athletics: How much similarity is there between sailing and, say, table tennis? About as little as between calligraphy and heavy metal. Again, the analogy between art and athletics is inescapable, since athletics, too, subsumes a wide range of intrinsically disparate activities defying a common denominator.

I believe that the essence of art lies less in the intrinsic properties of art objects (in a broad sense) and more in the nature of that which it does for us. I am here to propose that the origin and the function of art are akin to the origin and function of athletics. But if the raison d'être of athletics (or at least its essential aspect) is to exercise the body, the heart, the lungs, and the muscles, then the raison d'être of art, or at least its essential aspect, is to exercise the mind, to exercise the brain with its numerous and diverse parts serving numerous and diverse perceptual and cognitive functions. I propose that the function of art in society is to provide exercise for the mind and for the senses, and by so doing to boost brainpower in an open-ended way, not linked to any particular practical task. In this scheme of things, art and music are not mere, intrinsically frivolous markers of mental fitness, but they are in fact the critical tools of attaining and maintaining mental fitness. To those who may challenge this view by asking why a dedicated form of mental exercise is needed when we are constantly engaged in "real-life" utilitarian, mentally demanding activities, I will say that such activities are usually quite parochial and repetitive, constrained by the boundaries of one's professional and social roles in society. By contrast, art may have evolved as a more universal, more efficient, better-rounded, less parochial, and less vocation-bound way of exercising the mind, the senses, and the brain. In some sense, the proliferation of art forms in culture may have anticipated and prefigured the notion of a "cognitive workout circuit." Admittedly, this is a conjecture in need of further exploration, but it is a plausible one.

Art, like athletics, does not serve any particular, narrow survival function at any given time. This is precisely what frees it from the oppressive "I have to do it" bitter-pill quality and imbues it with the pleasant aura of a pursuit freely chosen, rather than an unavoidable, obligatory activity. People partake of the artistic and athletic avocations because they want to, not because they have to—a critical difference between vocation and avocation. But hidden underneath the pleasurable, inviting wrapping are the powerful tools of biological and cognitive self-improvement. For those who choose to indulge themselves, art and athletics stand apart from other human activities because of the combination of overt allure and tacit utility they offer.

The notion of art as a brain booster has already infiltrated the public consciousness or at least the public subconsciousness. Parents play Mozart to their infants (or even fetuses) in the expectation that this will foster their cognitive development. And an association between scientific or political genius and artistic avocation is also well known: think of Einstein's violin and Churchill's palette.

My former student Beth Neiman has made an interesting personal observation. Since she took up piano lessons a few months ago, she noticed an overall increase in her general sharpness and lucidity even in cognitive tasks far removed from music. The effect is most pronounced immediately after a music lesson—a cognitive workout of sorts. This effect is akin to the famous "Mozart effect": after listening to classical music you feel sharper in every respect. Evidently, many outstanding intellectuals, such as Einstein and Churchill, were tacitly (or possibly even overtly, but nobody asked) aware of this phenomenon.

Aging and Cognitive Fitness

Encouraged by the growing body of scientific evidence and by the thought that cognitive fitness has been with us in various guises for centuries and even for millennia, I felt that we were ready to develop our own cognitive enhancement program, one of the first of its kind. A handful of memory-enhancement programs have been described in scientific literature and have reported modest success. This was encouraging, but I felt that there was a huge reservoir of neuropsychological knowledge as of yet untapped and begging to be parlayed into practical applications.

In keeping with America's demographic changes, the face of my clinical practice was also changing. Increasingly, I was seeing men and women in their sixties, seventies, and eighties, some retired, others still active, all concerned with subtle or not so subtle signs of cognitive decline. In an insidious way, their angst resonated with my own. Most often, their profound concern was memory, but memory is a complex function and the introspective sense of "memory decline" may hide many things, including those having in reality very little to do with actual memory. In fact, the word *memory* is often used by the general public so broadly, synonymously with *cognition,* that a complaint about "memory impairment" carries relatively little information. Others were concerned about their absentmindedness, or about their inability to make decisions, or about their newly-acquired short temper.

In most cases we conducted a neuropsychological evaluation, very systematically and methodically measuring language, various forms of attention and memory, problem-solving, and other functions. Generally, human introspection into one's mental world is far less precise than people tend to believe it to be, and the last thing I was prepared to do was to take my patients' self-diagnoses on face value. The clinical analogy I like to draw—slightly crude, perhaps, but fundamentally accurate—is between a neuropsychologist and a dentist. When a patient complains of a toothache, a novice dentist may X-ray only the part of the palate where it hurts. But a seasoned dentist will X-ray the whole palate and is likely to discover an offending abscess in a totally different part of the mouth—the "referred-pain" phenomenon.

Often the neuropsychological evaluations turned up subtle signs of cognitive decline, but sometimes there were none. While some of our visitors clearly suffered from identifiable forms of early dementia, or at least mild cognitive impairment (MCI), many others did not and continued to live active, productive lives.

But they all complained of cognitive decline. Even when they showed no evidence of cognitive impairment on our tests, we couldn't ignore our patients' complaints. No matter how sensitive our tests are, they can miss a subtle cognitive change, particularly in a very bright individual. Generally, we do not know what the person was like five, ten, or twenty years ago. All we know is what we see now, at the time of the evaluation, and all we can do is compare our patient's performance to the supposed "norms," the reference data describing the typical performance of other individuals of comparable age, education, and other demographic characteristics. But what if our patient was not typical to begin with? What if he or she was exceptionally bright and gifted? Then despite a genuine cognitive decline, possibly even a significant one, the patient may still compare favorably with the typical population sample. I call it the "Einstein phenomenon." Albert Einstein with a twenty-IQ-point loss would still measure at a far higher intelligence level than your average Joe Blow, but Einstein himself would feel the difference.

What do you tell such people? While some of them come without any particular sense of concern merely because their doctors sent them, many others come on their own, driven by an inner sense of intense urgency and anxiety. They need more than a diagnosis; they need help. It was primarily for these people—bright, aging, concerned, and motivated to act on their concerns—that with the help of my assistants Peter Lang, Dmitri Bougakov, Lalita Krishnamurthy, Michael Zimmerman, Eric Rosenwinkel, and Jacqui Barnett, I developed our cognitive-enhancement program. In the design of the program, we quite intentionally emulated the traditional health club, the gym. We felt that the analogy with a known and established entity would make our cognitive-fitness program more understandable to the general public and would resonate with something already accepted by it. In scientific parlance such resonance is called "face validity." In a health club you encounter a number of machines, each designed to exercise a particular muscle group or physiological system. In our cognitive enhancement program you encounter a number of computer exercises, each designed to train a particular aspect of your mind. This means that instead of intimidating steel contraptions, barbells, dumbbells, and other typical health club paraphernalia, our cognitive fitness center has a lot of computers.

Earlier in the book, we talked about the multitude of complex functions of the mind. For each of them we have attempted to have a cognitive exercise (and often more than one) in our cognitive gym. We have designated separate exercises, or more often whole collections of exercises, for various aspects of memory, attention, language, reasoning, problem-solving, and so on. Of course, these are very broad categories, each subsuming a number of specific mental functions. For instance, attention is a broad category and one can distinguish between sustained attention, divided attention, and so on. Likewise, memory is a broad category and one can distinguish between verbal memory, visual memory for objects, visual memory for spatial configurations, and so on. Problem-solving may be essentially spatial, or verbal, or it may involve extrapolating things in time. We have attempted to address as many of these specific aspects of cognition as possible with special exercises.

To mention a few, an exercise designed to train sustained attention will challenge you by "wearing you down" with a long sequence of stimuli on the screen, each requiring a different response. By contrast, an exercise designed to train divided attention will have you respond to different events happening at the same time in different parts of the screen. An exercise designed to train planning abilities will have you navigate a route in a field where certain moves are permitted and others are not, and it is up to your ingenuity to figure out which are which. Now if the rules of the game change as you go, you will be forced to play catch-up with the changes, which will challenge and strain your mental flexibility to boot. And so on. Each exercise allows different difficulty levels, which are increased as progress is being made.

The exercises have the appearance of computer puzzle games,[2]

[2]In the world of computer game design a distinction is made between "action games," where the heroes slay dragons, and "puzzle games" and "strategy games," which are essentially computerized brainteasers.

but they have been carefully chosen to train particular aspects of the mind in a highly selective, targeted fashion. Because of the highly computerized (as opposed to paper-and-pencil) nature of our program, it would be difficult to include specific examples in this book, but I hope you get the idea.

In a health club you will be taken through the paces by a personal fitness trainer. In our cognitive-enhancement center you will also be met by a personal cognitive-fitness trainer, who will supervise and direct your cognitive exercises. The trainer will also turn your computers on and off and switch from one exercise to the next—a particularly helpful feature of the program for participants afflicted with computerphobia. The supervision is close but unobtrusive, and as a result the computerphobia in those afflicted is overcome and forgotten rather quickly.

Before starting the program, a baseline evaluation is conducted to identify the participant's cognitive strengths and cognitive weaknesses. We have enough cognitive exercises at our disposal to customize the workout program to the needs of each individual. If a specific profile emerges, we can decide whether we want to focus on the person's weaknesses, or to offer a broad, all-inclusive "cognitive cocktail."

Often we focus on the participant's weak area. This is sometimes met with the client's consternation: "Why do I have to do what is difficult for me, when I could be doing all these other, easier things?!" But if our theory is correct and cognitive exercise improves the function of the responsible brain structures, then the emphasis on the weak areas of cognition is the logical thing to do, the same way a golfer trying to reduce his handicap practices the elements of the game that are his weakest.

This method stands in sharp contrast to the philosophy once followed in the cognitive rehabilitation of patients recovering from stroke or from head injury. There an attempt has traditionally been made to teach the patient how to bypass, to circumvent, the impaired function rather than to improve it. But in physical medicine, like in brain science, there has been a growing appreciation of the body's natural plasticity and a consequent paradigm shift in the basic tenets of rehabilitation. Our ambitious philosophy of meeting the client's cognitive weaknesses head-on, as opposed to circumventing them, is consonant with the paradigm shift taking place in physical medicine and is inspired by the same recent findings of neuroscience.

It used to be the basic tenet of physical rehabilitation that when a patient loses the ability to use an arm or a leg as the result of a stroke or other debilitating neurological illness, you deal with this by teaching the patient how to use the spared limb to perform the functions previously controlled by the damaged one. But recently, owing to a large extent to the work of Edward Taub at the University of Alabama in Birmingham, a radically different, bold, and audacious approach has been developed. Instead of emphasizing the spared limb in the rehabilitation process, you immobilize it by literally putting a strap around it and encourage the patient to use the limp, presumably useless limb to do what it had been doing before the injury. This seemingly unrealistic activity has been amazingly successful in many cases. Evidently, putting the disabled limb to use promotes the development of new neural pathways and perhaps even the proliferation of new neurons in the brain areas damaged by the stroke. Or this may even spur other, spared parts of the brain (usually adjacent to the damaged ones) to take over the neural control of the disabled limb.

But let us go back to our program. As with physical fitness, you need to go to the gym regularly, and so the participants are usually encouraged to come to our cognitive enhancement program two or three times a week, one hour at a time. Each session consists of about a half dozen cognitive exercises, the composition of which may vary from session to session. At the inception of the program we planned to have individual sessions, a personal cognitive-fitness trainer working one-on-one with a program participant. But an interesting "couples" phenomenon emerged over time: a husband and wife, or two friends coming together. Then the session becomes an intimate family affair, each spouse sitting in front of his or her own computer terminal, each engaged in her/his own exercise program. They work together yet independently, with the trainer hovering over both of them.

A health club is for physical exercise; a cognitive-enhancement center is for cognitive exercise. But in addition to these explicit, main objectives, each helps fulfill a number of ancillary social needs for its members. As time went on, it became clear to us that our clients were developing personal bonds with the trainers and that for many of them it was an important ingredient of the total experience. Just the fact of going somewhere and interacting with other people also seemed to serve an important social function.

It was very interesting to watch how these personal agendas formed and evolved in our program participants. Some continued to come to the sessions like on a mission, doing the exercises, monitoring their progress without any social distractions. Others, by contrast, seemed to also come for the companionship of the personal cognitive-fitness trainers, all of them young and personable people, evidently filling the personal and social voids in some of our program participants' lives.

Usually a blend of cognitive and social interests developed, turning our clients, even the initially skeptical ones, into true devotees of the program. Even though the motivation behind the program design was cognitive, with time I have come to appreciate the therapeutic, albeit ancillary, value of the social context and recognize it as an important ingredient of success. Virtually without exception, all our clients enjoy the exercises and often linger past the end of the hour-long sessions. Many of them tend to become quite competitive with themselves, keeping track of their performance from session to session, making sure that continuous progress is taking place, and getting mad at themselves (despite the reassuring protestation of my staff) when the progress was only modest.

It seemed to me that this resurrection of competitive spirit in my aging clients, many of them years or even decades past the stage in lives when they had to compete for anything, had in itself a powerful therapeutic, invigorating, and rejuvenating effect. It was almost a magical feeling to see how life was pouring back into our aging clients. As it grows and matures, the program is benefiting from the feedback and reflections of our "students." These students are as diverse as New York City itself.

Diverse Like New York City

Louise is a retired writer and editor, seventy-two years old. Even though she lives in a posh, antiseptic part of Manhattan's Upper East Side, to me Louise is a quintessential "downtown" New Yorker: bright, blunt, always reveling in crossing all the *t*'s and dotting all the *i*'s, not bothering to leave much to imagination. She has lived a bohemian, freewheeling lifestyle often decades ahead of the prevailing social mores of the times.

While in her early seventies, Louise became concerned about her failing memory and attention. She was particularly frightened by her occasional episodes of confusion, like when she would take dirty dishes to the bedroom instead of the kitchen, leave her stove turned on, or, embarrassingly, forget to flush the toilet. In her own mind, Louise concluded that she was suffering from Alzheimer's disease and that the end of her mental, if not physical, life was near. A feeling of doom and helplessness was setting in, and she became increasingly hopeless about her future.

Louise contacted a famous New York neurologist who sent her to me. Her MRI and her neuropsychological evaluation were essentially normal. A few abnormal findings were equivocal and vastly out of proportion with Louise's concerns, but I could not discount the possibility of the "Einstein phenomenon" in this case. I told her about our program and she embraced it enthusiastically, becoming one of its first participants.

Louise stayed with the program for a few years, in the course of which a dramatic transformation took place. Gradually, the sense of doom had disappeared and gave way to the sense of (for lack of a better term) cognitive empowerment. While acknowledging that her memory was still bad, Louise shifted her focus from what she could not do to what she could. Through her work in the program, she gradually discovered all the things that she *could* do and concluded that these were aplenty.

After a few years in the program, Louise was ready for bigger and better things. She began taking college-level courses at a respectable local university. Even though she stopped attending our cognitive enhancement program, Louise kept in touch, calling me periodically to proudly report on her college work and to nag me occasionally with frustrations that her grades were only in the middle of her class and not at the top. We both agreed, however, that years after her resignation to the prospects of Alzheimer's disease (albeit imagined rather than real), her being able to hold her own in a class of students who could easily be her grandchildren was not so bad. After all, she felt confident enough to go to school again, after a fifty-six-year-long hiatus, and was getting A's.

I saw Louise at a book-signing party of my previous book, and she proudly announced that she had just received her bachelor's degree and was working toward a master's degree in social work. Having exorcised her own demons, she was ready to help others. In her conversations with me, Louise highlighted the difference the cognitive-enhancement program has made in her life. She felt that after regular mental exercises, her memory had shown a dramatic turnaround, and her mind was getting sharper. Louise has regained the confidence and competence she was afraid had been gone forever, and it gave her a new life. Participation in the cognitive-enhancement program relieved Louise's fear of Alzheimer's disease. It has taught her that exercising the brain is just as important as exercising the body to keep it in good shape. In her own words, the experience "jump-started" her to feeling empowered again and enabled her to "take control" of her memory, attention, and reasoning, and gave her a renewed sense of self-confidence.

Louise's transformation seemed nothing short of miraculous; but I do not believe in miracles. What, then, was at the heart of her "miraculous" story? I, like Louise, believe that our cognitive exercises had a direct effect on her cognitive functions. But I also believe that the very fact of being able to engage in the vigorous and rigorous cognitive activity that our twice-weekly workouts offered liberated Louise from her sense of doom and resignation. Before she joined our program, she had been in a state of "learned cognitive helplessness," and now she was free of it. The theme of empowerment, of being reconnected with cognitive *abilities,* rather than *disabilities,* is a common theme that runs through the accounts offered by many of our program participants about how they experience the effects of our program on them.

Just as the program gave Louise the confidence to take college courses after a fifty-six-year hiatus, it helped Elena to continue in her beloved profession, acting. Despite her age of eighty-two, Elena continues to enjoy a busy life on the stage. Yet she felt that her acting career was being jeopardized by her increasing difficulties with memorizing her lines—an essential skill for an actress. Elena is petite, with an irreverent, quick wit and a razor-sharp tongue. She is not intimidated by authority and occasionally calls me on the carpet, when she feels that I give a hard time to her personal cognitive-fitness trainer.

Elena contacted me on the recommendation of a friend, who heard about our program. At the time she was anxious and depressed about her ever-growing memory problems. While she could joke about her "senior moments" in social situations, in her heart she knew that it was no joking matter. Elena was increasingly overcome with the feeling of helplessness on a professional level. She seemed unable to retain even a moderate number of lines in the play in which she had just been cast. Even when Elena thought that she had finally committed them to memory, retrieval was still a problem at times. "[The words seemed to have] fallen out of my head," she would say.

In the early sessions of the program Elena found herself talking and trying to engage her trainer in conversation. It appeared that she was not eager to get to the computer and the solitary task at hand. By her own account, she felt that she had no patience. She felt that she "had no focus and no real concentration, two skills that are essential to being a good actor." But with time Elena found herself drawn into the program and increasingly enjoying the exercises. With time, Elena's attitude toward her cognitive lapses underwent drastic transformation. She no longer accepted her "senior moments" as a normal, inevitable part of getting older. She became more hopeful about the condition of her memory, and the cognitive exercises in our program became her way of trying to improve it. And some time ago Elena reported to me, happily, that she had just finished the run of a play "with my lines engraved in my memory, at least for now." Now, two and a half years into the program, Elena came to accept that there "would be no magical morning when I would waken to find my memory completely restored." But she feels that her head is clearer and that she is able "to unlock short-term memory more often than ever before," albeit not in a sustained manner. At about the time when Elena began to work with us, she also joined a study at a major medical center where she receives a complete battery of tests every two years. When Elena was tested recently for the two-year follow-up she showed neither gain nor loss. This, Elena and her doctors at the medical center agreed, is "essentially a gain, since at my age a loss would be likely."

The case of the retired physician, Dr. A., who is ninety years old, is particularly interesting, since it contains a bit of a clinical puzzle. A cultivated man, highly successful, proud and demanding of himself, he developed hydrocephalus. This condition, characterized by impaired drainage of cerebrospinal fluid (CSF) in the brain, is not an uncommon source of dementia in the elderly. To help the CSF drainage, Dr. A. had a thin tube ("shunt") surgically inserted, whose purpose is to drain the excess cerebrospinal fluid from the brain to the abdominal cavity, where it is absorbed. As often happens, the shunt had to be readjusted ("revised") a month later.

Dr. A. came to my office accompanied by his wife, also a retired professional. The two were obviously very close and the wife was getting increasingly concerned about her husband's cognition. Both highly educated people, they were guardedly inquisitive about our program. In the end they decided that Dr. A. would join the program and see what happened. Before he began to attend the cognitive enhancement sessions, we evaluated his cognition with a series of neuropsychological tests, as we usually do. These tests serve an important function of providing a frame of reference, a baseline with which to compare any future change.

Approximately three months after he started the program we reevaluated Dr. A. with a series of neuropsychological tests. His performance clearly improved across the board: memory, attention, and other functions. Such reevaluations at regular intervals are also extremely important, since they provide a precise, objective, and quantitative measure of the program participant's progress, or lack thereof. But we are always mindful of the fact that our neuropsychological tests and our cognitive exercises are useful only to the extent that they tell us something about our program participants' cognitive functioning in real life. No matter how sophisticated our tests are, they give us only very approximate and very imprecise impression of cognitive performance outside our office. If for no other reason, this is so because real-life circumstances, demands, and contexts are too individual, diverse, and varied to allow any meaningful standardization. This is why we ask our program participants and their family members to share with us directly their impressions about any impact the program may have on their real-life performance. Ultimately, this is what matters most. And so, about three months later I asked Dr. A. to comment on any perceived change in his cognition and addressed the same question to his wife.

Dr. A. felt that the twice-weekly sessions had definitely improved his recent memory. As a result, he felt more connected with daily events and activities. He was better able to remember what he was doing during the day and the day before. The occasions of meeting and talking with friends and relatives were better retained in his memory, as were his emotions at the time.

His wife also felt that Dr. A. had shown definite improvement and attributed his progress to the program. When Dr. A. started the program, his wife was really concerned that he was developing "senile dementia at a slow pace." But now she felt that his recent memory loss, although still present, appeared to have leveled off and that his concentration was much better. Most important, she felt that her husband's apathy, which had been growing at an alarming rate, was now gone and that he was much more like his former self—"judgmental, capable of enjoying concerts, theater . . ." She was overjoyed that Dr. A. was again reading a great deal and has been playing the piano much more often for brief periods, exhibiting a good "classical music memory."

They both emphasized that the context in which the therapy was taking place seemed to be as important as the therapy itself. "I still know that my principal and outstanding advantage in all this is that my wife is at my side and eternally supportive and helpful and loving," said Dr. A. And his wife confirmed that Dr. A. continued to rely a great deal on her memory, planning, and thinking things through—which seems to have been a longstanding feature of their interaction.

But could it be that our program was "a red herring" in this case? After all, Dr. A. had his shunt inserted, following which an MRI showed some diminution in the size of his lateral ventricles.

So, judging by the MRI, it appeared that the shunt placed in his brain a few months earlier was working and that the radiological symptoms of hydrocephalus had subsided. This had to have a therapeutic effect on his cognition as well. But the effects of the shunt are usually evident within a few weeks after insertion with subsequent stabilization of cognition. Dr. A. had only *started* our program two months after the shunt, and the baseline neuropsychological evaluation was conducted at that time. The cognitive improvement evident in our tests and reported by Dr. A. and his wife was noted relative to that baseline and therefore was not likely to have been due to the shunt. It had to have something to do with our program!

For some of our program participants the motivation is prevention rather than treatment. One of them is Paul, a successful international entrepreneur. Bright and dynamic, Paul looks and acts much younger than his passport age of sixty-five. Nor does he have any inkling of an impending cognitive decline. Paul is a voracious reader, and every time he comes to my office he carries a new book in his hand. Nonetheless, Paul decided to join our program as a way of protecting and prolonging his cognitive sharpness as he ages. Paul feels that since beginning the cognitive-exercise program he experiences less anxiety, is more analytical, and has better focus. His ability to deal with complex new material has also improved. As an example, Paul brought up his recent experience with listening to a Schoenberg composition. He was pleased to notice that not only did he not become "defensive with the first few atonal sounds but welcomed their challenge, remaining focused on the notes and analyzing the juxtaposition of the notes, the chords, and in general the composition itself."

Paul has also noted with satisfaction that in uncomfortable situations he has become more conscious of listening, analyzing, and reflecting before responding. He finds himself to be less impulsive and rushing in his reactions than in the past. Paul also noted "better attention when confronted with challenging problems or technical and tedious writings." And then Paul made a very interesting and subtle distinction. He did not necessarily feel that his scope of intellectual capacity has increased, but rather that he learned how to utilize it better. This, too, may be a useful byproduct of cognitive exercise.

Memory may be the most frequently voiced concern, but often enough our program participants notice changes in other mental functions. One of them is a semiretired physician, Dr. B. He found out about our program from a friend, also a program participant. Dr. B. has become so enthusiastic about the program that he has been urging his own patients to join it. He feels that there has been an improvement in his memory, but he is particularly pleased to feel an increased "capacity to see alternative paths to take in a course of action" and an improved ability to "plan better for the future" and "to learn from mistakes." Dr. B. also reports having "more of a sense of empowerment in my daily activities." This account clearly points to an improvement in executive functions, the functions of the frontal lobes.

A common theme running across the participants' comments is that the program demystifies cognition for them in their own minds. It is no longer one diffuse, undifferentiated, all-embracing, fog-like "thing" that can be lost in its entirety, as if through an act of cruel black magic. By engaging in various exercises, doing better on some and worse on others, program participants learn to "anatomize" their own cognition. Invariably, certain cognitive functions are better preserved than others, and this gives them a sense of reassurance and control. The mere fact of learning how to delineate the scope of one's weaknesses, as well as the scope of one's strengths, has a powerful therapeutic effect. The positive effect of cognitive engagement is often apparent also by its absence. Many participants notice that when they miss a few consecutive weeks in the program they find themselves "duller."

Being introduced to, and becoming comfortable with, the

digital world has been an ancillary benefit of the program that a number of our "students" have enjoyed. Coming into the program without any knowledge of computers and often with a hefty dose of computerphobia, they end up developing modest computer skills and comfort levels, and find all kinds of computer applications in their lives above and beyond our cognitive enhancement program.

People who find us and join our program come from different backgrounds, are driven by different concerns and anxieties, and are animated by different hopes and expectations. They are all welcome and we try to help them all. A program participant wrote this poignant verse:

Who is this person I hear when I hunt for a word?
My memory, once like a safe, its content held fast
Yet rendered at need, before the occasion was past,
Has lapses now, frustrating and absurd.

We do what we can to help people like him, and other people with less dramatic signs of decline, or even with no apparent signs of decline. Do our methods make a real impact, and if so, how do we know it? The standard neuropsychological tests, with which we reevaluate our program participants at set intervals, provide a partial answer to this question. In many instances definite improvement is noted. This is invariably a source of delight both for the client and for us. In other cases no improvement is evident. But in most of the cases showing no improvement there is no evidence of decline either. This is, on balance, also good news, since we are dealing with a population where the expected baseline without intervention is one of gradual decline. But the most important evidence of success, modest though it often is, is the impression of the program participants and their family members that they are retaining—or even regaining—their mental powers in real life.

In this pursuit we are guided by the famous adage attributed to Albert Einstein: "Nothing is more practical than a good theory." Growing understanding of the mechanisms of the brain's lifelong plasticity and of its capacity for rejuvenating itself throughout the life span will continue to increase our ability to extend and enrich the lives of the minds of real people in the real world.

EPILOGUE

THE PRICE OF WISDOM

I repeated the MRI two and a half years later. There was no "interval change," no suggestion of any progressive brain disease; but the punctate lesion that my friend Sandy Antin had declared an artifact was not there. Indeed, it must have been an artifact. So, with any luck, my aging brain will continue to serve me in good stead for the foreseeable future—exactly for how long, who knows.

Does this mean that I have attained wisdom? At least enough not to flatter myself that I have. But like all of you, I have amassed my share of neural pattern-recognition devices, enabling me to understand my world and to act in it with a good-enough degree of efficiency. The entirety of these patterns, and also of those patterns that could have become part of my mental arsenal but haven't, are the sum total of my life's experiences, my mental striving, and my mental slack. I remember myself aged six talking to a neighbor aged fifteen and trying to imagine how a person that old must feel. Today I am a generally satisfied owner of a fifty-eight-year-old brain, feeling fine and wondering, what it is like to be seventy, eighty, or ninety years old.

The idea of this book was prompted by my introspection into the changes attending the flow of seasons of my own mind. The purpose was to glean beyond the introspection and try to grasp the mechanisms behind the mental changes. To that end, I have endeavored to examine the seasons of human mind both in the cultural and in the neurobiological contexts, and to connect the two vantage points into a coherent "natural history of the mind" through the life span. The natural history narrated in this book is admittedly incomplete. For instance, we barely touched on our mind's moral and spiritual aspects, on how the moral and spiritual premises develop, and how they inform our mental life.

Despite these obvious omissions, as I am approaching the final stage of my inquiry, I find the result, on balance, satisfying, for I feel that the natural history of the mind that emerged from this inquiry makes sense, that it illuminates and informs my own introspection, certainly not completely but to a reasonable degree. The feeling, that "the seasons of our mind" are not all downhill and that some important mental gains are attained as we age, is grounded in neurobiological reality; it is not merely a desperate exercise in wishful thinking by an aging intellectual.

Has this reassurance completely removed the angst of my ripe middle age? Of course not. Do I regret that it hasn't? No, I don't, since in modest amounts such angst can be a great constructive force, motivating and mobilizing, reminding one that our time is finite and thus should not be squandered. But the two messages emanating loud and clear from my natural history of the mind are, on balance, reassuring.

The first message is that those of us whose mental lives have been both vigorous and rigorous approach their advanced years with a mighty coat of mental armor. This armor, a mental autopilot of sorts, will serve them in good stead during the final decades of their lives. This mental armor, the rich collection of pattern-recognizing attractors in the brain, is not an entitlement and its attainment in old age is not a foregone conclusion. It is a reward for the vigorous life of the mind in younger years.

We all hate clichés, often forgetting that what makes them clichés is that they are grounded in truth. Platitudes are boring not because they are wrong but because they are truistically self-evident. "The past is the best predictor of the future" is a well-worn cliché, but like most clichés it contains a huge dose of truth. It is true in history, in economics, in politics, and it is true in the lives of our brains and thus of our minds.

The seemingly effortless ability to "see through things" that, depending on its caliber and context, we call competence or expertise, or in rare instances wisdom, does not come by itself as an epiphany of maturity or as an entitlement of old age. It is the condensation of mental activities across years and decades of life. The scope and quality of one's mental lifetime will shape the quality of its final stages. "Wisdom begins in wonder," said Socrates. This is as true now as it was then, maybe more so.

Our journey through life is through the life of our minds. A life of the mind rich in experiences, faced with mental challenges frequently, diversely, and unabashed by them, rewards us with a generous arsenal of cognitive tools. These cognitive tools empower us mightily as we age and shield us against the effects of brain decay. Life is finite—we all know that—but we prepare the stage for the endgame by the lifelong totality of our experiences and endeavors. This is true for our bodies, and it is equally true for our minds.

The second message is that, while taking full advantage of mental autopilot, one must not allow oneself to be lulled by it. Regardless of one's age, one must continue to test one's mind and strive for new mental challenges. In these times of infatuation with physical fitness everybody has heard about the "runner's high," a surge of joy caused by physical exertion and physical accomplishment. But how many among us have experienced the feeling of a "thinker's high?" This feeling is dear to some scientists and artists. Not to all, mind you. Membership in a creative profession does not automatically mean the life of a creative mind. A famous chess player, an acquaintance of my parents when I was a little boy in Riga, once said: "Most people play chess with their hands, and only very few do it with their heads." Even the most exalted intellectual vocations hold out the seductive option of sluggish mental assembly line. Recognize it for what it is, and don't allow it to take over!

Setting aside experiencing it—how many members of the general public have even contemplated the possibility of having the feeling of a "thinker's high?" How many people realize that there is such a thing as mental exertion? And even when people do understand, how far beyond mere rhetoric does this understanding go? How many among us truly recognize that strenuous thinking is an activity in its own right, occurring in space and time? When I try to explain to people that for me the hours spent walking my dog is time not wasted but gained, because it is my "pure thinking" time, the time that allows me to do all kinds of things in my head, including writing this book, too often I have the feeling that people don't have the slightest clue of what I am talking about and probably believe that I am making it up to make my idle pastimes look respectable. With some people the notion of pure thought does not seem to cut muster, even when attached to the indisputably productive activity of walking a dog. But one should know better and one should listen to the poet:

My Mind to me a kingdom is,
Such present joys therein I find,
That it excels all other bliss
That earth affords or grows by kind.

As England was shaking off the last vestiges of dogma-dominated medieval slumber and firmly claiming its place in the blossoming of the Renaissance, the age of Shakespeare, Newton, and Elizabethan Enlightenment, this verse by Sir Edward Dyer (1543?–1607) was emblematic of the rediscovered fascination with lively pursuits of the mind. Today as then, those who enjoy such pursuits strengthen and protect their minds from decay.

Some people are physically vigorous, and this carries lifelong rewards. Others are physically lazy, and this, too, has lifelong consequences. Equally, some people crave mental challenges, and others regard them as hardships. Given the choice, they stay within a seductively cozy mental comfort zone without realizing that a mental comfort zone is a mental stagnation zone. "There is no expedient to which a man will not go to avoid the real labor of thinking," was Thomas Edison's nihilistic pronouncement. While it certainly does not apply to all, it does unfortunately apply to many, maybe even to most. Make sure *you* are not part of that mentally sluggish slice of humanity!

Just as physical laziness comes at a price, so, too, does mental laziness. Mental laziness in youth endangers your brain in old age. Remember William James's admonition about not squandering one's formative, "plastic" years. Those who take delight in mental challenges, and seek them out above and beyond the bare necessities of workaday existence, scaffold their minds and their brains with powerful protective gear, which will go a long way toward ensuring a sound and rich mental life well into old age.

But vigorous life of the mind should not come to a halt at any time. It can, and must, continue well into advanced age. The longer it goes on, the longer it will continue to bestow its own rewards in the form of stimulating various growth processes in the brain and by so doing protecting it from the effects of decay. The concept of lifelong mental fitness, with better odds for keeping a sound mind for life as its reward, should become part of popular culture. I believe it soon will.

The image of a sage is among the most revered in every culture. After a long infatuation with youth, aging is respected and admired again even in our impatient, conceited culture. Given the massive demographic trends, it better be! One is not born to be a sage—one becomes a sage as the reward for a long journey. The journey I speak of is a journey of the mind. It is the ascendancy to wisdom. Aging is the price of wisdom, but wisdom itself is priceless. To those to whom it comes, it comes as the golden age of the mind.

As for me, I may decide to have another MRI in a few years'

time. My ventricles may be a bit larger and my sulci may start to give. Another tiny area of hyperintensity may show up. But I will take all this in stride and in good humor. My head is full of attractors, and blissfully I will put my mental autopilots to good use. And by way of tweaking my mind with new mental challenges to keep churning out those new neurons in my aging brain and help prevent it from failing beyond hope and redemption, I have written this book. With any luck, there will be more to come.

ACKNOWLEDGMENTS

Several people helped make this book possible in a variety of ways. Michelle Tessler, my agent at Carlisle and Company, has my deepest gratitude for placing my proposal in the caring hands of Gotham Books. I could not have hoped for a better publisher. I thank my superb editor at Gotham, Brendan Cahill, and his assistant Patrick Mulligan, who guided me from the inception of the project to its finale in a thoughtful, patient, and constructive way. I thank Dmitri Bougakov for his extensive help with various technical aspects of book preparation and for his substantive comments on the manuscript. Peter Lang has been my right hand in the cognitive enhancement program described in the book. Richard Gallini contributed the illustrations for the book. Fiona Stevens, Kate Edgar, Sergey Knazev, Lalita Krishnamurthy, and Brendan Connors offered valuable advice and assistance at various stages of the project. I thank my patients and the cognitive enhancement program participants for the opportunity to work with them and by so doing gaining the insights and the experience that served as the foundation of the book—and for allowing me to tell their stories and to quote them. I thank "Steve's" father for the permission to write about his son. My students served as the captive audience before whom I rehearsed pieces of the book disguised as lectures. I thank them for their indulgence.

Finally, I thank my dog Brit, a most unlikely muse. As a boy growing up in the Baltic city of Riga (then part of the Soviet Union), I had two family dogs, and this formed a lifelong affection for canines. My father was an involuntary resident of Stalin's "hotel" called Gulag. My mother was fired from her college teaching job and became a factory laborer working brutally long shifts to support the family. The three of us, my dogs and I, spent time together in the communal apartment where we lived. By the age of three I had come to relate to them as close friends. As an adult, I always wanted a dog but was too busy, too protective of my freedom, my lifestyle too peripatetic. Finally, perhaps as part of the midlife jolt that prompted this book, I decided to get a dog. In time Brit, the bullmastiff puppy, turned into a majestic leonine creature of intimidating stature but good heart, noble disposition, and superior canine intelligence, my friend and companion.

I have lived on the edge of New York's Central Park for many years, yet never made much use of the proximity, except enjoying the views of the park's expansive lawns and lush treetops from my living-room windows. But now, forced to get up uncharacteristically early to walk Brit, I brought my handheld computer along and spent many hours in Central Park writing or thinking through the yet unwritten chapters while sipping espresso at the dog-friendly Sheep Meadow Cafe or resting by Bethesda Fountain, with Brit dozing off at my feet, nagging me for a biscuit, trying to devour the bench on which I was sitting, and basically not contributing anything of substance to the creative process. These quiet very early mornings in the park brought with them focus and clarity of thought and offered a precious, if temporary, escape from frenetic Manhattan life. In its essential features the book came together over the course of one summer, while walking the dog. The rest was easy: just putting it on paper.

The book is dedicated to my generation, the baby boomers, whose anxieties and hopes I understand and share.

CHAPTER NOTES

Introduction

Tolstoy's *Anna Karenina*: Tolstoy, L. (2003). *Anna Karenina.* New York: Barnes & Noble Classics.

A book, an intellectual memoir of sorts: Goldberg, E. (2001). *The Executive Brain: Frontal Lobes and the Civilized Mind.* Oxford; New York: Oxford University Press; paperback 2002.

1. The Life of Your Brain

Mind-body dualism: Damasio, A. (1994). *Descartes' Error; Emotion, Reason, and the Human Brain.* New York: Putnam Publishing Group; Pinker, S. (2002). *The Blank Slate: The Modern Denial of Human Nature.* New York: Viking.; Koestler, A. (1967). *The Ghost in the Machine.* London,: Hutchinson; Goldberg, E. (2001). *The Executive Brain: Frontal Lobes and the Civilized Mind.* New York: Oxford University Press; paperback 2002.

Herbert Simon's work: Simon, H. A. (1996). *The Sciences of the Artificial* (3rd ed.). Cambridge, MASS.: MIT Press.

Attractors: Grossberg, S. (Ed.). (1988). *Neural Networks and Natural Intelligence.* Cambridge: MIT Press.

"The alarm has just rung, rudely assaulting your brain stem, your thalamus and your auditory cortex." More about the basics of human neuroanatomy and neuropsychology in: Kolb, B., & Whishaw, I. Q. (1996). *Fundamentals of Human Neuropsychology* (4th ed.). New York: W. H. Freeman.

Dysfunction of the right hemisphere: Rourke, B. P. (1989). *Nonverbal Learning Disabilities: The Syndrome and the Model.* New York: The Guilford Press.

Hippocampi in Alzheimer's disease: de Leon, M. J., Convit, A., George, A. E., Golomb, J., de Santi, S., Tarshish, C., et al. (1996). In vivo structural studies of the hippocampus in normal aging and in incipient Alzheimer's disease. *Annals of NY Acad Sci,* 777, 1–13.

Prefrontal cortex function and dysfunction: Goldberg, E. (2001 paperback 2002). *The Executive Brain: Frontal Lobes and the Civilized Mind.* Oxford; New York: Oxford University Press.

ADHD: Barkley, R. A. (1997). *ADHD and the Nature of Self-Control.* New York: The Guilford Press.

2. Seasons of the Brain

Brain development: Brown, M., Keynes, R., Lumsden, A. (2002). *The Developing Brain.* New York: Oxford University Press: Harvey, D.S., et al., Eds. (2000). *Development of the Nervous System.* New York: Academic Press; Carpenter, M. B., & Parent, A. (1995). *Carpenter's Human Neuroanatomy* (9th ed.). Baltimore: Lippincot, Williams & Wilkins.

Neural Darwinism: Edelman, G. M. (1987). *Neural Darwinism: The Theory of Neuronal Group Selection.* New York: Basic Books.

Brain aging: Raz, N. (2000). Aging of the brain and its impact on cognitive performance: integration of structural and functional findings. In F. Craik & T. Salthouse (Eds.), *The Handbook of Aging and Cognition* (2nd ed., p. 1). Mahwah, NJ: Lawrence Erlbaum Associates.

Naftali Raz on Aesculapius: *Ibid.*

"Evolution and dissolution": Jackson, H. (1884). Evolution and dissolution of the nervous system. *Cronian Lecture. Selected papers, 2.*

Cognitive aging: F. Craik & T. Salthouse, Eds. (2000). *The Handbook of Aging and Cognition* (2nd ed.). Mahwah, NJ: Lawrence Erlbaum Associates; D. C. Park & N. Schwartz, Eds. (2000). *Cognitive Aging: A Primer.* Philadelphia: Psychology Press.

Prefrontal cortex: Goldberg, E. (2001; paperback 2002). *The Executive Brain: Frontal Lobes and the Civilized Mind.* Oxford; New York: Oxford University Press.

Aged individuals' performance in real-life situations: Park, D., & Gutchess, A. (2000). Cognitive aging and everyday life. In D. C. Park & N. Schwarz (Eds.), *Cognitive Aging: A Primer.* Philadelphia: Psychology Press.

Cognitive expertise in aging: Raz, N. (2000). Aging of the brain and its impact on cognitive performance: integration of structural and functional findings. In F. Craik & T. Salthouse (Eds.), *The Handbook of Aging and Cognition* (2nd ed., p. 47). Mahwah, NJ: Lawrence Erlbaum Associates.

3. Aging and Powerful Minds in History

Goethe's *Faust*: Goethe, J. W. v. (1994). *Faust. Parts 1 and 2.* New York: Continuum.

Gaudi: Constantino, M. (1993). *Gaudi.* Secaucus, NJ: Chartwell Books, Inc.

Grandma Moses: Nikola-Lisa, W. (2000). *The Year with Grandma Moses* (1st ed.). New York: Henry Holt.

Wiener's writings: Wiener, N. (1948). *Cybernetics.* New York: J. Wiley; Wiener, N. (1964). *God and Golem, Inc.: A Comment on Certain Points Where Cybernetics Impinges on Religion.* Cambridge: MIT Press.

Chillida: Ezquiaga, M. (2001). *Museo Chillida-Leku* (2nd ed.): Chillida-Leku S. L.; Weber, S., Hammacher, A. M., Trier, E., & de Baranano, K. (2002). *Chillida.* Kèunzelsau: Swiridoff.

Much of the discussion about Willem de Kooning is based on works by Sally Yard and Edvard Lieber: Yard, S. (1997). *Willem de Kooning.* New York: Rizzoli; Lieber, E. (2000). *Willem de Kooning: Reflections in the Studio.* New York: Abrams.

Hyden Herrera's account: Yard, S. (1997). *Willem de Kooning.* New York: Rizzoli.

"A finished painting is a reminder of what not to do tomorrow": Quoted after Lieber: Lieber, E. (2000). *Willem de Kooning: Reflections in the Studio.* New York: Abrams.

"I'm back to a full palette with off-toned colors. Before it was about knowing what I didn't know. Now, it's about not knowing what I know": Quoted after Lieber: *Ibid.*

"Style is a fraud . . . To desire to make a style is an apology for one's anxiety": Quoted after Yard: Yard, S. (1997). *Willem de Kooning.* New York: Rizzoli.

" 'The rhythms are more deliberate, meditated even, and the space more open . . . a new order prevails, a new calm . . . de Kooning has purified his stroke, and what had been quintessentially sensuous is rendered immaterial, ethereal, a veiled tracing of its physical origins,' wrote David Rosand.": *Ibid* p. 104.

" 'de Kooning, who has never strayed far from nature for long, is closer to it now than ever,' wrote Vivien Raynor.": Raynor, V. (June 13, 2002). *The New York Times,* p. A18.

Plato on wisdom: Plato (2000). *The Republic.* Mineola, NY: Dover Publications. Quoted after Csikszentmihalyi, M., & Rathunde, K. (1990). The psychology of wisdom: an evolutionary interpretation. In: R. Sternberg (Ed.) *Wisdom: Its Nature, Origins, and Development.* (pp. 25–51). New York: Cambridge University Press, p. 33.

Newton's, Kant's, and Faraday's memory loss with age: Sacks, O. (2003). *Personal communication to E. Goldberg.*

Shannon's Alzheimer's disease: Johnson, G. (February 27, 2001). Mathematician dies at 84. *The New York Times.*

Reagan's familial risk factors: Altman, L. K. (June 15, 2004). A recollection of early questions about Reagan's health. *The New York Times,* pp. F5, 10.

Hitler's Parkinson's disease: Irving, D. (1983). *The Secret Diaries of Hitler's Doctor.* New York: William Morrow.

Parkinson's disease and dementia: Aarsland, D., Andersen, K., Larsen, J. P., Lolk, A., & Kragh-Sorensen, P. (2003). Prevalence and characteristics of dementia in Parkinson disease: an 8-year prospective study. *Archives of Neurol, 60*(3), 387–392.

Hitler's mental condition: Speer, A. (1981). *Inside the Third Reich: Memoirs.* New York: Collier Books.

Hitler's and Stalin's memory decline: Neumayr, A. (1995). *Dictators in the Mirror of Medicine.* Medi-Ed Press.

Much of the discussion about Stalin's mental condition is based on: Conquest, R. (1992). *Stalin: Breaker of Nations.* New York: Penguin; Brent, J., & Naumov, V. P. (2003). *Stalin's Last Crime: The Plot against the Jewish Doctors, 1948–1953* (1st ed.). New York: HarperCollins.

Stalin's senility, including the accounts by Krushchev, Djilas and Vinogradov: Montefiore, S.S. (2004). *Stalin: The Court of the Red Tsar.* New York: Alfred A. Knopf; Neumayr, A. (1995). *Dictators in the Mirror of Medicine,* Medi-Ed Press.

Lenin's strokes: Clark, R. W. (1988). *Lenin, the Man Behind the Mask.* London; Boston: Faber and Faber; Volkogonov, D. A., & Shukman, H. (1994). *Lenin: A New Biography.* New York: Free Press.

Lenin's possible syphilis: Chivers, C. J. (June 22, 2004). A retrospective diagnosis says Lenin had syphilis. *The New York Times,* p. F3; Golding, M. (July 18, 2004). "Psychiatrists Say Lenin Died of Syphilis." Reuters.

Mao's physical and mental condition: Li, Z. (1994). *The Private Life of Chairman Mao: The Memoirs of Mao's Personal Physician.* New York: Random House.

Dementia in ALS: Strong, M. J. (2001). Progress in clinical neurosciences: the evidence for ALS as a multisystems disorder of limited phenotypic expression. *Canadian J Neurol Sci, 28*(4), 283–298.

Hitler and Stalin remaining in control till the end: Bullock, A. (1993). *Hitler and Stalin: Parallel Lives*. New York: Vintage Books.

Woodrow Wilson's last years in the White House: Smith, G. (1982). *When the Cheering Stopped: The Last Years of Woodrow Wilson*. Alexandria, VA: Time-Life Books.

FDR's mental powers and decision-making abilities: Jenkins, R. (2001). *Churchill: A Biography* (1st ed.) (p. 774). New York: Farrar, Straus and Giroux.

FDR's "new disinclination to apply himself to serious business" quoted from: *Ibid* (p. 774).

Churchill's mental lapses and strokes: Danchev, A., & Todman, D. (Eds.). (2001). *War Diaries 1939–1945: The Diaries of Field Marshal Lord Alanbrooke*: Orion Publishing Co.; Jenkins, R. (2001). *Churchill: A Biography* (1st ed.). New York: Farrar, Straus and Giroux.

Churchill *"gloriously unfit for office"*: Quoted from Jenkins, R. (2001). *Churchill: A Biography* (1st ed.) (p. 845). New York: Farrar, Straus and Giroux.

Margaret Thatcher's strokes: BBC. (2002). *Thatcher suffers 'minor stroke'*. Retrieved January 26, 2002, from http://news.bbc.co.uk./1/hi/uk/1783722.stm; Lyall, S. (June 8, 2004). Thatcher's tribute was waiting: *The New York Times,* p. A23.

Brezhnev's senility: Volkogonov, D. (1998). *The Rise and Fall of the Soviet Empire: Political Leaders from Lenin to Gorbachev.* New York: HarperCollins.

Successful aging: Rowe, J., & Kahn, R. (1998). *Successful Aging*. New York: Random House.

4. Wisdom Throughout Civilizations

For a comprehensive and insightful review of psychological research on wisdom see: Sternberg, R. (Ed.). (1990). *Wisdom: Its Nature, Origins, and Development*. New York: Cambridge University Press.

Peter Thompson's interviews: Thompson, P. (2003). *Wisdom: The Hard-Won Gift*. Adelaide: Griffin Press.

Mihaly Csikszentmihalyi and Kevin Rathunde on wisdom: Csikszentmihalyi, M., & Rathunde, K. (1990). The psychology of wisdom: an evolutionary interpretation. In R. Sternberg (Ed.), *Wisdom: Its Nature, Origins, and Development* (pp. 25–51). New York: Cambridge University Press: pp. 25–51.

Sophocles on wisdom: Sophocles (2003). *Antigone*. New York: Oxford University Press. Quoted after Ibid.

Bible on wisdom: *The Holy Bible: Proverbs, 4:*7. (2002). Grand Rapids, MI: Zondervan. Quoted after Ibid.

James Birren and Laurel Fisher on early mentions of wisdom: Birren, J., & Fisher, L. (1990). The elements of wisdom: overview and integration. In R. Sternberg (Ed.), *Wisdom: Its Nature, Origins, and Development* (pp. 317–332). New York: Cambridge University Press: p. 319.

Paul Baltes and Jacqui Smith on wisdom, wisdom tree, and as expert knowledge: Baltes, P., & Smith, J. (1990). Toward a psychology of wisdom and its ontogenesis. In R. Sternberg (Ed.), *Wisdom: Its Nature, Origins, and Development* (pp. 87–120). New York: Cambridge University Press.

The Wisdom Tree: Ibid.; Sears, E. (1986). *Ages of Man: Medieval Interpretations of the Life Cycle.* Princeton: Princeton University Press.

Seven Pillars of Wisdom: Lawrence, T. E. (1991). *Seven Pillars of Wisdom: A Triumph* (1st Anchor Books ed.). New York: Anchor Books.

"To understand wisdom fully and correctly probably requires more wisdom than any of us have": Quoted from Sternberg, R, (1990). Understanding wisdom. In R. Sternberg (Ed.), *Wisdom: Its Nature, Origins, and Development* (p. 3). New York: Cambridge University Press.

Robert Sternberg on wisdom and creativity: Sternberg, R. (1985). Implicit theories of intelligence, creativity and wisdom. *Journal of Personality and Social Psychology, 49*(3), 607–627.

Daniel Robinson on wisdom: Robinson, D. (1990). Wisdom through ages. In R. Sternberg (Ed.), *Wisdom: Its Nature, Origins, and Development* (p. 21). New York: Cambridge University Press.

J.F.C. Fuller on genius: Quoted after Bose, P. (2003). *Alexander the Great's Art of Strategy.* New York: Gotham Books.

William Wordsworth on greatness and originality: Quoted after Greenberg, N. (2003). "The Executive Brain: Frontal Lobes and the Civilized Mind," by Elkhonon Goldberg. *Human Nature Review, 3*, 422–431. Original source: Wordsworth, W. (1969). "William Wordsworth's Letter to Lady Beaumont, 21 May 1807." In E. de Selincourt (Ed.), *Letters of William and Dorothy Wordsworth Vol. 2.*

Carl Rogers on creativity: Quoted from Rogers, C. R. (1961). *On Becoming a Person: A Therapist's View of Psychotherapy.* Boston: Houghton Mifflin.

Robert Sternberg on competence and wisdom: Sternberg, R. (1985). Implicit theories of intelligence, creativity and wisdom. *Journal of Personality and Social Psychology, 49*(3), 607–627.

Themistocles "was greater in genius than in character.": Quoted after Bose, P. (2003). *Alexander the Great's Art of Strategy* (p. 81). New York: Gotham Books.

Popular perception of wisdom, creativity and intelligence: Sternberg, R. (1990). Wisdom and its relations to intelligence and creativity. In R. Sternberg (Ed.), *Wisdom: Its Nature, Origins, and Development* (p. 145). New York: Cambridge University Press.

Human attributes, their desirability, and age: Heckhausen, J., Dixon, R., & Baltes, P. (1989). Gains and losses in development throughout adulthood as percieved by different adult age groups. *Developmental Psychology, 25*, 109–121.

Marion Perlmutter on wisdom and advanced age: Orwoll, L., & Perlmutter, M. (1990). The study of wise persons: integrating a personality perspective. In R. Sternberg (Ed.), *Wisdom: Its Nature, Origins, and Development* (pp. 160–180). New York: Cambridge University Press.

5. Pattern Power

"Phyletic" knowledge: Fuster, J. M. (2003). *Cortex and Mind: Unifying Cognition*. New York: Oxford University Press.

Relatively ready-to-use, but requiring early environmental honing neural networks: Hubel, D. H., & Wiesel, T. N. (1963). Receptive fields of cells in striate cortex of very young, visually inexperienced kittens. *J Neurophysiol, 26*, 994–1002; Hubel, D. H., & Wiesel, T. N. (1979). Brain mechanisms of vision. *Sci Am, 241*(3), 150–162.

Primate cultures: Wrangham, R. W., & Chicago Academy of Sciences. (1994). *Chimpanzee Cultures*. Cambridge: Published by Harvard University Press in cooperation with the Chicago Academy of Sciences.

Primate language: Savage-Rumbaugh, S., Shanker, S. G., & Taylor, T. J. (2001). *Apes, Language, and the Human Mind*. New York: Oxford University Press.

Language learning: Pinker, S. (1994). *The Language Instinct* (1st ed.). New York: W. Morrow and Co.

Eskimo language: Pullum, G. K. (1991). *The Great Eskimo Vocabulary Hoax and Other Irreverent Essays on the Study of Language*. Chicago: University of Chicago Press.

Click languages: Stephenson, J. (2000). *The Language of the Land: Living among The Hadzabe in Africa* (1st ed.). New York: St. Martin's Press.

Whistling language: Meyer, J. (2004). Bioacoustics of human whistled languages: an alternative approach to the cognitive processes of language. *Anais da Academia Brasiliera de Ciências, 76*(2), 406–412.

Complexity of an organism's behavior and environment: Simon, H. A. (1996). *The Sciences of the Artificial* (3rd ed.). Cambridge: MIT Press.

Emergent complexity: Wolfram, S. (2002). *A New Kind of Science*. Champaign, IL: Wolfram Media, Inc.

Cortical representation of language: Martin, A., Haxby, J. V., Lalonde, F. M., Wiggs, C. L., & Ungerleider, L. G. (1995). Discrete cortical regions associated with knowledge of color and knowledge of action. *Science, 270*(5233), 102–105; Martin, A., Wiggs, C. L., Ungerleider, L. G., & Haxby, J. V. (1996). Neural correlates of category-specific knowledge. *Nature, 379*(6566), 649–652. Also, for more detailed discussion on the topic see: Goldberg, E. (1989). Gradiental approach to neocortical functional organization. *J Clin Exp Neuropsychol, 11*(4), 489–517; Goldberg, E. (1990). Higher Cortical Functions in Humans: The Gradiental Approach. In E. Goldberg (Ed.), *Contemporary Neuropsychology and the Legacy of Luria* (pp. 229–276). Hillsdale, NJ: Lawrence Erlbaum Associates; Goldberg, E. (2001). *The Executive Brain: Frontal Lobes and the Civilized Mind*. Oxford; New York: Oxford University Press.

Vygotsy's writings: Vygotsky, L. S. (1962). *Thought and Language*. Cambridge, Mass.: MIT Press; Rieber, R. W., Robinson, D. K., & Bruner, J. S. (Eds.). (2004). *The Essential Vygotsky*. Kluger Academic/Plenum.

Luria's life in science: Luria, A. R., Cole, M., & Cole, S. (1979). *The Making of Mind: A Personal Account of Soviet Psychology*. Cambridge: Harvard University Press; Goldberg, E. (1990); Tribute to Aleksandr Romanovich Luria (1902–1977). In E. Goldberg (Ed.), *Contemporary Neuropsychology and the Legacy of Luria* (pp. 1–9). Hillsdale, NJ: Lawrence Erlbaum Associates; Moskovich, L., Bougakov, D., DeFina, P., & Goldberg, E. (2002). A. R. Luria: Pursuing Neuropsychology in a Swiftly Changing Society. In A. Stringer, E. Cooley & A. L. Christensen (Eds.), *Patways to Prominence in Neuropsychology*. New York: Psychology Press.

Luria's neuropsychological research: Luria, A. R. (1970). *Traumatic Aphasia*: The Hague: Mouton; Luria, A. R. (1966). *Higher Cortical Functions in Man*. New York: Basic Books.

Emergent cortical organization: Goldberg, E. (1989). Gradiental approach to neocortical functional organization. *J Clin Exp Neuropsychol, 11*(4), 489–517; Goldberg, E. (1990); Higher Cortical Functions in Humans: The Gradiental Approach. In E. Goldberg (Ed.), *Contemporary Neuropsychology and the Legacy of Luria* (pp. 229–276). Hillsdale, NJ: Lawrence Erlbaum Associates.

Gradiental principle of functional cortical organization: *Ibid.*

6. Adventures on Memory Lane

Role of neocortex in memory: Goldberg, E., & Barr, W. (1992). Selective knowledge loss in activational and representational amnesias. In L. Squire & N. Butters (Eds.), *Neuropsychology of Memory* (pp. 72–80). New York: The Guilford Press; Fuster, J. M. (2003). *Cortex and Mind: Unifying Cognition*. New York: Oxford University Press.

Role of other structures in memory: Goldberg, E., & Barr, W. (1992). Selective knowledge loss in activational and representational amnesias. In L. Squire & N. Butters (Eds.), *Neuropsychology of Memory* (pp. 72–80). New York: The Guilford Press.

Memory without forgetting: Luria, A. R. (1968). *The Mind of a Mnemonist: A Little Book About a Vast Memory*. New York: Basic Books.

Causes of amnesia: Squire, L., & Schacter, D. (Eds.). (2002). *Neuropsychology of Memory* (3rd ed.). New York: The Guilford Press.

Perception and memory share the same cortical territory: Kosslyn, S. M., Thompson, W. L., & Alpert, N. M. (1997). Neural systems shared by visual imagery and visual perception: a positron emission tomography study. *Neuroimage, 6*(4), 320–334; Kosslyn, S. M., Thompson, W. L., Kim, I. J., & Alpert, N. M. (1995). Topographical representations of mental images in primary visual cortex. *Nature, 378*(6556), 496–498.

Reverberating loops: Hebb, D. O. (1949). *The Organization of Behavior: A Neuropsychological Theory*. New York: Wiley.

Long-term potentiation: Bashir, Z. I., Bortolotto, Z. A., Davies, C. H., Berretta, N., Irving, A. J., Seal, A. J., et al. (1993). Induction of LTP in the hippocampus needs synaptic activation of glutamate metabotropic receptors. *Nature, 363*(6427), 347–350.

Role of hippocampi in memory: Maviel, T. et al (2004) Sites of neocortical reorganization critical for remote spatial memory. *Science,* 305, 96–99; Remondes, M., & Schuman, E. M. (2004). Role for a cortical input to hippocampal area CA1 in the consolidation of a long-term memory. *Nature,* 431 (7009), 699–703.

Electric shock in studies of memory formation: Glickman, S. E. (1961). Perseverative neural processes and consolidation of the memory trace. *Psychol Bull, 58*, 218–233; McGaugh, J. L. (1972). The search for the memory trace. *Ann NY Acad Sci, 193*, 112–123.

Permastore: Bahrick, H. P. (1984). Semantic memory content in permastore: fifty years of memory for Spanish learned in school. *J Experimental Psychol Gen, 113*(1), 1–29.

Distribution of memories in permastore: *Ibid*.

Retrograde amnesia: Goldberg, E., Antin, S. P., Blider, R. M., Jr., Gerstman, L. J., Hughes, J. E., & Mattis, S. (1981). Retrograde amnesia: possible role of mesencephalic reticular activation in long-term memory. *Science, 213*(4514), 1392–1394; Goldberg, E., & Barr, W. (1992). Selective knowledge loss in activational and representational amnesias. In L. Squire & N. Butters (Eds.), *Neuropsychology of Memory* (pp. 72–80). New York: The Guilford Press.

Temporal gradient of retrograde amnesia: Barr, W. B., Goldberg, E., Wasserstein, J., & Novelly, R. A. (1990). Retrograde amnesia following unilateral temporal lobectomy. *Neuropsychologia, 28*(3), 243–255.

7. Memories That Do Not Fade

Neural Darwinism: Edelman, G. M. (1987). *Neural Darwinism: The Theory of Neuronal Group Selection*. New York: Basic Books.

Procedural and declarative memory: Cohen, N. J., & Squire, L. R. (1980). Preserved learning and retention of pattern-analyzing skill in amnesia: dissociation of knowing how and knowing that. *Science, 210*(4466), 207–210.

Episodic and semantic memory. Tulving, E. (1983). *Elements of Episodic Memory*. New York: Oxford University Press.

More on "Steve's" case in: Goldberg, E., Antin, S. P., Bilder, R. M., Jr., Gerstman, L. J., Hughes, J. E., & Mattis, S. (1981). Retrograde amnesia: possible role of mesencephalic reticular activation in long-term memory. *Science, 213*(4514), 1392–1394. As the title of the paper suggests, the damage responsible for memory impairment was in the ventral midbrain, this further supporting the role of the brain stem in memory. See also E. Goldberg, J. Hughes, S. Mattis & S. Antin. (1982). Isolated retrograde amnesia: Different etiologies, same mechanisms? *Cortex*, 18, 459–462.

Generic and specific memory: Goldberg, E., & Barr, W. (1992). Selective Knowledge Loss in Activational and Representational Amnesias. In L. Squire & N. Butters (Eds.), *Neuropsychology of Memory* (pp. 72–80). New York: The Guilford Press; Goldberg, E., & Barr, W. (2003). Knowledge Systems and Material-Specific Memory Deficits. In J. H. Byrne (Ed.), *Learning and Memory*. New York: Macmillan Reference.

Generic and specific memory in retrograde amneisa: Barr, W. B., Goldberg, E., Wasserstein, J., & Novelly, R. A. (1990). Retrograde amnesia following unilateral temporal lobectomy. *Neuropsychologia, 28*(3), 243–255.

Generic memories are committed to long-term storage more rapidly: Goldberg, E., & Barr, W. (1992). Selective Knowledge Loss in Activational and Representational Amnesias. In L. Squire & N. Butters (Eds.), *Neuropsychology of Memory* (pp. 72–80). New York: The Guilford Press; Maviel, T., Durkin, T. P., Menzaghi, F., & Bontempi, B. (2004). Sites of neocortical reorganization critical for remote spatial memory. *Science, 305*(5680), 96–99.

To learn more about brain plasticity see this excellent book: Schwartz, J., & Begley, S. (2002). *The Mind and the Brain: Neuroplasticity and the Power of Mental Force*. New York: Regan Books.

Pattern expansion in monkeys. Wang, X., Merzenich, M. M., Sameshima, K., & Jenkins, W. M. (1995). Remodelling of hand representation in adult cortex determined by timing of tactile stimulation. *Nature, 378*(6552), 71–75.

Pattern expansion in humans: Pascual-Leone, A., & Torres, F. (1993). Plasticity of the sensorimotor cortex representation of the reading finger in Braille readers. *Brain, 116(Pt 1)*, 39–52; Elbert, T., Pantev, C., Wienbruch, C., Rockstroh, B., & Taub, E. (1995). Increased cortical representation of the fingers of the left hand in string players. *Science, 270*(5234), 305–307.

Pattern expansion and mental clarity in old age: Golden, D. (1994, July). "Building a Better Brain." *Life*, 62–70.

Brain metabolic requirements in learning: Haier, R. J., Siegel, B. V., Jr., MacLachlan, A., Soderling, E., Lottenberg, S., & Buchsbaum, M. S. (1992). Regional glucose metabolic changes after learning a complex visuospatial/motor task: a positron emission tomographic study. *Brain Res, 570*(1–2), 134–143.

Cortical activation changes with task familiarization: Dobbins, I. G., Schnyer, D. M., Verfaellie, M., & Schacter, D. L. (2004). Cortical activity reductions during repetition priming can result from rapid response learning. *Nature, 428*(6980), 316–319.

Prions and memory: Wickelgren, I. (2004). Neuroscience. Long-term memory: a positive role for a prion? *Science, 303*(5654), 28–29.

Memory reconsolidation: Lee, J. L., Everitt, B. J., & Thomas, K. L. (2004). Independent cellular processes for hippocampal memory consolidation and reconsolidation. *Science, 304*(5672), 839–843.

Attractor, attractor state, basin of attraction: Grossberg., S. (Ed.). (1988). *Neural Networks and Natural Intelligence*. Cambridge: MIT Press; Fuster, J. M. (2003). *Cortex and Mind: Unifying Cognition*. Oxford; New York: Oxford University Press.

Degeneracy: Edelman, G. M. (1987). *Neural Darwinism: The Theory of Neuronal Group Selection*. New York: Basic Books.

Attractors in mathematics: Professor Alan Snyder (2003), personal communication to E. Goldberg.

Attractors as memories: Hopfield, J. J. (1982). Neural networks and physical systems with emergent collective computational abilities. *Proceedings of Natl Acad Sci USA, 79*(8), 2554–2558.

Attractor-like circuits in the brain: Cossart, R., Aronov, D., & Yuste, R. (2003). Attractor dynamics of networks UP states in the neocortex. *Nature, 423*(6937), 283–288.

Attractor-like effects in classifications: Freedman, D. J., Riesenhuber, M., Poggio, T., & Miller, E. K. (2001). Categorical representation of visual stimuli in the primate prefrontal cortex. *Science, 291*(5502), 312–316.

Adaptive resonance theory (ART): Grossberg, S. (1987). Competitive learning: from interactive activation to adaptive resonance. *Cognitive Science, 11*, 23–63; Grossberg, S. (Ed.). (1988). *Neural Networks and Natural Intelligence*. Cambridge: MIT Press.

Goldberg on modularity: Goldberg, E. (1995). Rise and fall of modular orthodoxy. *J Clin Exp Neuropsychol, 17*(2), 193–208.

8. Memories, Patterns, and the Machinery of Wisdom

Intuitive decision making: Simon, H. A. (1996). *The Sciences of the Artificial* (3rd ed.). Cambridge: MIT Press.

Phyletic knowledge: Fuster, J. M. (2003). *Cortex and Mind: Unifying Cognition*. Oxford; New York: Oxford University Press.

Pattern recognition in problem-solving: Simon, H. A. (1996). *The Sciences of the Artificial* (3rd ed.). Cambridge: MIT Press.

Harold Bloom on genius: Bloom, H. (2002). *Genius: A Mosaic of One Hundred Exemplary Creative Minds*. New York: Warner Books.

Charles Murray on human accomplishment: Murray, C. A. (2003). *Human Accomplishment: The Pursuit of Excellence in the Arts and Sciences, 800 BC to 1950*. New York: HarperCollins.

Peter Thompson on wisdom: Thompson, P. (2003). *Wisdom: The Hard-Won Gift*. Adelaide: Griffin Press.

Allan Snyder's TMS experiments: Snyder, A. W., Mulcahy, E., Taylor, J. L., Mitchell, D. J., Sachdev, P., & Gandevia, S. C. (2003). Savant-like skills exposed in normal people by suppressing left fronto-temporal lobe. *Journal of Integrative Neuroscience*, 2:2.

William James on habits: *The Principles of Psychology* (Vol. 1). New York: Dover.

Descriptive knowledge, veridical knowledge: Goldberg, E., Harner, R., Lovell, M., Podell, K., & Riggio, S. (1994). Cognitive bias, functional cortical geometry, and the frontal lobes: laterality, sex, and handedness. *Journal of Cognitive Neuroscience, 6*(3), 276–296; Goldberg, E., & Podell,

K. (2000). Adaptive decision making, ecological validity, and the frontal lobes. *J Clin Exp Neuropsychol, 22*(1), 56–68.

Prescriptive knowledge, actor-centered knowledge: *Ibid.*

9. "Up-Front" Decision-Making

Frontal lobotomy: Valenstein, E. (1986). *The Great and Desperate Cures.* New York: Basic Books.

Prefrontal regions in charge of overall decision making vs. task subcomponents: Koechlin, E., Basso, G., Pietrini, P., Panzer, S., & Grafman, J. (1999). The role of the anterior prefrontal cortex in human cognition. *Nature, 399*(6732), 148–151.

Prefrontal cortical activation in rational problem solving: Kroger, J. K., Sabb, F. W., Fales, C. L., Bookheimer, S. Y., Cohen, M. S., & Holyoak, K. J. (2002). Recruitment of anterior dorsolateral prefrontal cortex in human reasoning: a parametric study of relational complexity. *Cereb Cortex, 12*(5), 477–485.

Prefrontal resources in inductive and deductive reasoning: Osherson, D., Perani, D., Cappa, S., Schnur, T., Grassi, F., & Fazio, F. (1998). Distinct brain loci in deductive versus probabilistic reasoning. *Neuropsychologia, 36*(4), 369–376.

Executive memories: Fuster, J. M. (2003). *Cortex and Mind: Unifying Cognition.* New York: Oxford University Press.

Frontal lobes serve as the repository of executive memories: *Ibid.*

Intentionality, ethical behavior, morality and empathy as subjects of cognitive neuroscience and experimental psychology: Goldberg, E. (2001; paperback 2002). *The Executive Brain: Frontal Lobes and the Civilized Mind.* New York: Oxford University Press.

Social neuroscience: Cacioppo, J. T. (2002). *Foundations in Social Neuroscience.* Cambridge: MIT Press.

Behavioral economics: Kahneman, D., & Tversky, A. (2000). *Choices, Values, and Frames.* New York: Cambridge University Press.

Neuroeconomics: Sanfey, A. G., Rilling, J. K., Aronson, J. A., Nystrom, L. E., & Cohen, J. D. (2003). The neural basis of economic decision-making in the ultimatum game. *Science, 300*(5626), 1755–1758.

Neuromarketing: Thompson, C. (October 26, 2003). "There Is as Sucker Born in Every Medial Prefrontal Cortex." *New York Times Magazine,* 54–57.

Functional neuroimaging and political commercials: Tierney, J. (April 20, 2004). "Using M.R.I.'s to See Politics on the Brain." *The New York Times,* pp. A1, A17.

Capital punishment in the mentally retarded: Beckman, M. (2004). Neuroscience. Crime, culpability, and the adolescent brain. *Science, 305*(5684), 596–599.

Figure 12: Adapted from Vogeley, K., Podell, K., Kukolja, J., Schilbach, L., Goldberg, E., Zilles, K., et al. (2003). Recruitment of the Left Prefrontal Cortex in Preference-Based Decisions in Males (fMRI Study). Paper presented at the Ninth Annual Meeting of the Organization for Human Brain Mapping, New York.

Figure 13: Adapted from Brodmann, K. (1909). *Vergleichende Lokalisationslehre der Grosshinrinde in ihren Prinzipien dargestellt auf Grund des Zellenbaues.* Leipzig: Barth.

Creation of Adam spin-off: Goldberg, E. (2001; paperback 2002). *The Executive Brain: Frontal Lobes and the Civilized Mind.* New York: Oxford University Press.

Frontal lobes in empathy: Singer, T., Seymour, B., O'Doherty, J., Kaube, H., Dolan, R. J., & Frith, C. D. (2004). Empathy for pain involves the affective but not sensory components of pain. *Science, 303*(5661), 1157–1162.

Frontal lobes in theory of mind: Fletcher, P. C., Happe, F., Frith, U., Baker, S. C., Dolan, R. J., Frackowiak, R. S., et al. (1995). Other minds in the brain: a functional imaging study of "theory of mind" in story comprehension. *Cognition, 57*(2), 109–128.; Stone, V. E., Baron-Cohen, S., & Knight, R. T. (1998). Frontal lobe contributions to theory of mind. *J Cogn Neurosci, 10*(5), 640–656.

Lack of insight following frontal damage: Goldberg, E. (2001; paperback 2002). *The Executive Brain: Frontal Lobes and the Civilized Mind.* New York: Oxford University Press.

Criminality, antisocial personality, and impulsive aggression linked to prefrontal dysfunction: Raine, A., Buchsbaum, M., & LaCasse, L. (1997). Brain abnormalities in murderers indicated by positron emission tomography. *Biol Psychiatry, 42*(6), 495–508; Raine, A., Lencz, T., Bihrle, S., LaCasse, L., & Colletti, P. (2000): Reduced prefrontal gray matter volume and reduced autonomic activity in antisocial personality disorder. *Arch Gen Psychiatry, 57*(2), 119–127; discussion 128–119.

Causal learning and the frontal lobes: Turner, D. C., Aitken, M. R., Shanks, D. R., Sahakian, B. J., Robbins, T. W., Schwarzbauer, C., et al. (2004). The role of the lateral frontal cortex in causal associative learning: exploring preventative and super-learning. *Cereb Cortex, 14*(8), 872–880.

"If . . . then . . ." structures in complex language: Fitch, W. T., & Hauser, M. D. (2004). Computational constraints on syntactic processing in a nonhuman primate. *Science, 303*, 377–380.

Experience of regret: Camille, N., Coricelli, G., Sallet, J., Pradat-Diehl, P., Duhamel, J. R., & Sirigu, A. (2004). The involvement of the orbitofrontal cortex in the experience of regret. *Science, 304*(5674), 1167–1170.

Myelinization of prefrontal pathways: Goldberg, E. (2001; paperback 2002). *The Executive Brain: Frontal Lobes and the Civilized Mind.* New York: Oxford University Press.

Spindle cells: Allman, J. M., Hakeem, A., Erwin, J. M., Nimchinsky, E., & Hof, P. (2001). The anterior cingulate cortex. The evolution of an interface between emotion and cognition. *Ann N Y Acad Sci, 935,* 107–117.

Emotional intelligence: Goleman, D. (1995). *Emotional Intelligence.* New York: Bantam Books.

Memory for concepts of actions: Fuster, J. M. (2003). *Cortex and Mind: Unifying Cognition.* New York: Oxford University Press.

High-functioning elderly and physiologically active frontal lobes: Cabeza, R., Anderson, N. D., Locantore, J. K., & McIntosh, A. R. (2002). Aging gracefully: compensatory brain activity in high-performing older adults. *Neuroimage, 17*(3), 1394–1402; Rosen, A. C., Prull, M. W., O'Hara, R., Race, E. A., Desmond, J. E., Glover, G. H., et al. (2002). Variable effects of aging on frontal lobe contributions to memory. *Neuroreport, 13*(18), 2425–2428.

Executive talent: Goldberg, E. (January 2004). Train the Gifted. *Harvard Business Review,* 31.

IQ following frontal-lobe damage: Goldberg, E. (2001; paperback 2002). *The Executive Brain: Frontal Lobes and the Civilized Mind.* New York: Oxford University Press.

10. Novelty, Routines, and the Two Sides of the Brain

Grossberg's "adaptive resonance": Grossberg, S. (1987). Competitive learning: from interactive activation to adaptive resonance. *Cognitive Science, 11,* 23–63.

Corpus callosum, commisures and cross-talk between the hemispheres: Kolb, B., & Whishaw, I. Q. (1996). *Fundamentals of Human Neuropsychology* (4th ed.). New York: W. H. Freeman.

Aphasia and left- vs. right-hemispheric damage: Luria, A. R. (1966). *Higher Cortical Functions in Man.* New York: Basic Books.

Aphasia and left- vs. right-hemispheric damage in children: Bates, E.

(1999). Plasticity, localization and language development. In S. Broman & J. Fletcher (Eds.), *The Changing Nervous System: Neurobehavioral Consequences of Early Brain Disorders* (pp. 214–253). New York: Oxford University Press.

Electric stimulation of the left temporal lobe and verbal hallucinatory-like experiences: Ojemann, G. A. (1983). Brain organization for language from the perspective of electrical stimulation mapping. *Behavioral and Brain Sciences, 6*, 189–230.

Auditory hallucinations in schizophrenia: Nasrallah, H. S. (Ed.). (1991). *Handbook of Schizophrenia*. New York; Amsterdam: Elsevier.

"Pathological" left-handedness: Orsini, D. L., & Satz, P. (1986). A syndrome of pathological left-handedness. Correlates of early left hemisphere injury. *Arch Neurol, 43*(4), 333–337.

Hyperphasia and Williams' syndrome: Personal communication from Dr. Oliver Sacks to E. Goldberg.

Damage to the right hemisphere, prosopagnosia and amusia: Luria, A. R. (1966). *Higher Cortical Functions in Man*. New York: Basic Books.

Larger *planum temporalem* and *frontal operculum* in the left hemisphere: Geschwind, N., & Levitsky, W. (1968). Human brain: left-right asymmetries in temporal speech region. *Science, 161*(837), 186–187.

Great apes and brain "language structures": LeMay, M., & Geschwind, N. (1975). Hemispheric differences in the brains of great apes. *Brain Behav Evol, 11*(1), 48–52; Gannon, P. J., Holloway, R. L., Broadfield, D. C., & Braun, A. R. (1998). Asymmetry of chimpanzee planum temporale: humanlike pattern of Wernicke's brain language area homolog. *Science, 279*(5348), 220–222.

Australopithecus and brain asymmetry: LeMay, M. (1976). Morphological cerebral asymmetries of modern man, fossil man, and nonhuman primate. *Ann NY Acad Sci, 280*, 349–366.

Yakovlevian torque: Geschwind, N., & Galaburda, A. M. (1985). Cerebral lateralization. Biological mechanisms, associations, and pathology. *Arch Neurol, 42*(5), 422–459.

Differences in the size of the *planum temporale* and *frontal operculum*: Geschwind, N., & Levitsky, W. (1968). Human brain: left-right asymmetries in temporal speech region. *Science, 161*(837), 186–187.

Brain asymmetries and cortical thicknesses: Galaburda, A. M., LeMay, M., Kemper, T. L., & Geschwind, N. (1978). Right-left asymmetrics in the brain. *Science, 199*(4331), 852–856; Diamond, M. C., Dowling, G. A., & Johnson, R. E. (1981). Morphologic cerebral cortical asymmetry in male and female rats. *Exp Neurol, 71*(2), 261–268; Diamond, M. C. (1985). Rat forebrain morphology: Right-left; male-female; young-old;

enriched-impoverished. In S. D. Glick (Ed.), *Cerebral laterality in nonhuman species*. New York: Academic Press.

Brain asymmetries and spindle cells: Blakeslee, S. (December 9, 2003). "Humanity? Maybe It's in the Wiring." *The New York Times,* pp. F1, 6.

Brain asymmetries and neurotransmitter pathways: Glick, S. D., Ross, D. A., & Hough, L. B. (1982). Lateral asymmetry of neurotransmitters in human brain. *Brain Res, 234*(1), 53–63; Sholl, S. A., & Kim, K. L. (1990). Androgen receptors are differentially distributed between right and left cerebral hemispheres of the fetal male rhesus monkey. *Brain Res, 516*(1), 122–126; Ebstein, R. P., Novick, O., Umansky, R., Priel, B., Osher, Y., Blaine, D., et al. (1996). Dopamine D4 receptor (D4DR) exon III polymorphism associated with the human personality trait of novelty seeking. *Nat Genet, 12*(1), 78–80.

Left-right hippocampal asymmetries and NMDA receptors: Kawakami, R., Shinohara, Y., Kato, Y., Sugiyama, H., Shigemoto, R., & Ito, I. (2003). Asymmetrical allocation of NMDA receptor epsilon2 subunits in hippocampal circuitry. *Science, 300*(5621), 990–994.

Brain asymmetries in fruit flies: Isabel, G., Pascual, A., & Preat, T. (2004). Exclusive consolidated memory phases in drosophila. *Science, 304*(5673), 1024–1027.

Left hemisphere as the repository of compressed knowledge: Goldberg, E., & Costa, L. D. (1981). Hemisphere differences in the acquisition and use of descriptive systems. *Brain Lang, 14*(1), 144–173.

Hemispheric specialization in left-handers: Rasmussen, T., & Milner, B. (1977). The role of early left-brain injury in determining lateralization of cerebral speech functions. *Ann NY Acad Sci, 299,* 355–369.

11. Brain Duality in Action

Novelty-routinization theory: Goldberg, E., & Costa, L. D. (1981). Hemispheric differences in the acquisition and use of descriptive systems. *Brain Lang, 14*(1), 144–173.

Familiar vs. "twisted" verbal task and cerebral hemispheres: *Ibid.*

Familiar vs. unfamiliar visual task and cerebral hemispheres: Marzi, C. A., & Berlucchi, G. (1977). Right visual field superiority for accuracy of recognition of famous faces in normals. *Neuropsychologia, 15*(6), 751–756.

Functional neuroimaging and right-to-left "cognitive center of gravity" transfer: Haier, R. J., Siegel, B. V., Jr., MacLachlan, A., Soderling, E., Lottenberg, S., & Buchsbaum, M. S. (1992). Regional glucose metabolic changes after learning a complex visuospatial/motor task: a positron emission tomographic study. *Brain Res, 570*(1–2), 134–143; Raichle, M. E., Fiez, J. A., Videen, T. O., MacLeod, A. M., Pardo, J. V., Fox, P. T., et al. (1994). Practice-related changes in human brain functional anatomy during nonmotor learning. *Cereb Cortex, 4*(1), 8–26; Gold, J. M., Berman, K. F., Randolph, C., Goldberg, T. E., & Weinberger, D. (1996). PET validation of a novel prefrontal task: Delayed response alteration. *Neuropsychology, 10,* 3–10; Tulving, E., Markowitsch, H. J., Craik, F. E., Habib, R., & Houle, S. (1996). Novelty and familiarity activations in PET studies of memory encoding and retrieval. *Cereb Cortex, 6*(1), 71–79; Berns, G. S., Cohen, J. D., & Mintun, M. A. (1997). Brain regions responsive to novelty in the absence of awareness. *Science, 276*(5316), 1272–1275; Martin, A., Wiggs, C. L., & Weisberg, J. (1997). Modulation of human medial temporal lobe activity by form, meaning, and experience. *Hippocampus,* 7(6), 587–593; Shadmehr, R., & Holcomb, H. H. (1997). Neural correlates of motor memory consolidation. *Science, 277*(5327), 821–825; Henson, R., Shallice, T., & Dolan, R. (2000). Neuroimaging evidence for dissociable forms of repetition priming. *Science, 287*(5456), 1269–1272.

Gamma EEG and cerebral hemispheres: Kamiya, Y., Aihara, M., Osada, M., Ono, C., Hatakeyama, K., Kanemura, H., et al. (2002). Electrophysiological study of lateralization in the frontal lobes. *Japanese Journal of Cognitive Neuroscience, 3:1,* 88–191.

Figure 14: *Ibid.*

Novice vs. expert musicians and cerebral hemispheres: Bever, T. G., & Chiarello, R. J. (1974). Cerebral dominance in musicians and nonmusicians. *Science, 185*(150), 537–539.

The role of the right hemisphere in language acquisition in children: For detailed review see: Goldberg, E., & Costa, L. D. (1981). Hemispheric differences in the acquisition and use of descriptive systems. *Brain Lang, 14*(1), 144–173; Bates, E. (1999). Plasticity, Localization and Language Development. In S. Broman & J. Fletcher (Eds.), *The Changing Nervous System: Neurobehavioral Consequences of Early Brain Disorders* (pp. 214–253). New York: Oxford University Press; Bates, E., & Roe, K. (2001). Language Development in Children with Unilateral Brain Injury." In C. A. Nelson & M. Luciana (Eds.), *Handbook of Developmental Cognitive Neuroscience.* Cambridge: MIT Press.

Damage to the right hemisphere and language in adults: Basser, L. S. (1962). Hemiplegia of early onset and the faculty of speech with special reference to the effects of hemispherectomy. *Brain, 85*, 427–460; Dennis, M., & Whitaker, H. A. (1976). Language acquisition following hemidecortication: linguistic superiority of the left over the right hemisphere. *Brain Lang, 3*(3), 404–433; Bates, E. (1999). Plasticity, Localization and Language Development. In S. Broman & J. Fletcher (Eds.), *The Changing Nervous System: Neurobehavioral Consequences of Early Brain Disorders* (pp. 214–253). New York: Oxford University Press.

The role of the right hemisphere in language and age: Ibid.

"Eureka!-like" insight in verbal puzzles: Jung-Beeman, M., Bowden, E. M., Haberman, J., Frymiare, J. L., Arambel-Liu, S., Greenblatt, R., et al. (2004). Neural activity when people solve verbal problems with insight. *PLoS Biol, 2*(4), E97.

The brain dynamics of the second language: Kim, K. H., Relkin, N. R., Lee, K. M., & Hirsch, J. (1997). Distinct cortical areas associated with native and second languages. *Nature, 388*(6638), 171–174; Lee, S., Yeon, E., Lee, D., & Jung, K. (2003). *Cortical Representations in Korean-English Bilinguals.* Ninth Annual Meeting of the Organization for Human Brain Mapping Conference, New York City; Mechelli, A., Noppeney, U., O'Doherty, J., Ashburner, J., & Price, C. (2003). A Voxel-Based Morphometry Study of Monolinguals, Early Bilinguals and Late Bilinguals. Ninth Annual Meeting of the Organization for Human Brain Mapping Conference, New York City; Meyer, M., Goddard, G., Simonotto, E., McNamara, A., Azuma, R., Flett, S., et al. (2003). Differential Brain Responses to L1 and L2 in Near-Native L2 Speakers. Ninth Annual Meeting of the Organization for Human Brain Mapping Conference, New York City.

Right-hemispheric stroke in a bilingual: Barbara Kapetanakes, personal communication to E. Goldberg.

Associative agnosia and damage to the left hemisphere: Goldberg, E. (1990). Associative agnosias and the functions of the left hemisphere. *J Clin Exp Neuropsychol, 12*(4), 467–484.

Ideational apraxia and damage to the left hemisphere: *Ibid.*

Byron Rourke's contribution to understanding right-hemispheric dysfunction: Rourke, B. P. (1989). *Nonverbal Learning Disabilities: The Syndrome and the Model.* New York: The Guilford Press.

Right-to-left shift of the "center of cognitive gravity" throughout life span: Cabeza, R., Grady, C. L., Nyberg, L., McIntosh, A. R., Tulving, E. Kapur, S., et al. (1997). Age-related differences in neural activity during memory encoding and retrieval: a positron emission tomography study. *J Neurosci, 17*(1), 391–400; Madden, D. J., Turkington, T. G., Provenzale, J. M., Denny, L. L., Hawk, T. C., Gottlob, L. R., et al. (1999). Adult age differences in the functional neuroanatomy of verbal recognition memory. *Hum Brain Map,* 7(2), 115–135; Aihara, M., Aoyagi, K., Goldberg, E., & Nakazawa, S. (2003). Age shifts frontal cortical control in a cognitive bias task from right to left: part I. Neuropsychological study. *Brain & Development,* 25, 555–559; Brown, T. T., Lugar, H. M., Coalson, R. S., Miezin, F. M., Petersen, S. E. & Schlaggar, B. L. (2004). Developmental changes in human cerebral functional organization for word generation. *Cerebral Cortex,* bhh129 (Electronic version).

Left prefrontal activation in older adults: Cabeza, R., Anderson, N. D., Locantore, J. K., & McIntosh, A. R. (2002). Aging gracefully: compensatory brain activity in high-performing older adults. *Neuroimage, 17*(3), 1394–1402.

Jason Brown and Joseph Jaffe on cerebral dominance: Brown, J. W., & Jaffe, J. (1975). Hypothesis on cerebral dominance. *Neuropsychologia, 13*(1), 107–110.

Functional neuroimaging, frontal lobes and task familiarity: Jahanshahi, M., Dirnberger, G., Fuller, R., & Frith, C. D. (2000). The role of the dorsolateral prefrontal cortex in random number generation: a study with positron emission tomography. *Neuroimage, 12*(6), 713–725; Reichle, E. D., Carpenter, P. A., & Just, M. A. (2000). The neural bases of strategy and skill in sentence-picture verification. *Cognit Psychol, 40*(4), 261–295.

Carlsson's experiment on low and high creativity: Carlsson, I., Wendt, P. E., & Risberg, J. (2000). On the neurobiology of creativity. Differences in frontal activity between high and low creative subjects. *Neuropsychologia, 38*(6), 873–885.

Creative people and increased right frontal activity: Martindale, C., & Hines, D. (1975). Creativity and cortical activation during creative, intellectual and EEG feedback tasks. *Biol Psychol, 3*(2), 91–100; Carlsson, I., Wendt, P. E., & Risberg, J. (2000). On the neurobiology of creativity. Differences in frontal activity between high and low creative subjects. *Neuropsychologia, 38*(6), 873–885.

12. Magellan on Prozac

Left-hemispheric damage and depression: Gainotti, G. (1972). Emotional behavior and hemispheric side of the lesion. *Cortex, 8*(1), 41–55; Narushima, K., Kosier, J. T., & Robinson, R. G. (2003). A reappraisal of poststroke depression, intra- and inter-hemispheric lesion location using meta-analysis. *J Neuropsychiatry Clin Neurosci, 15*(4), 422–430.

Right-hemispheric damage and mania or *belle indifference*: Goldstein, K.

(1939). *The Organism.* New York: American Books; Gainotti, G. (1972). Emotional behavior and hemispheric side of the lesion. *Cortex, 8*(1), 41–55.

Right-hemispheric damage and "anosognosia": Heilman, K., & Valenstein, E. (Eds.). (1993). *Clinical Neuropsychology.* New York: Oxford University Press.

Left hemineglect: *Ibid.*

"Alien hand" phenomenon: Goldberg, G., & Bloom, K. K. (1990). The alien hand sign. Localization, lateralization and recovery. *Am J Phys Med Rehabil*, 69(5), 228–238.

Left frontal lobe damage and depression: Robinson, R. G., Kubos, K. L., Starr, L. B., Rao, K., & Price, T. R. (1984). Mood disorders in stroke patients. Importance of location of lesion. *Brain 107* (Pt 1), 81–93; Davidson, R. (1995). Cerebral Assymetry, Emotion, and Affective Style. In R. Davidson & K. Hugdahl (Eds.), *Brain Assymetry* (pp. 361–388). Cambridge, MA: The MIT Press.

Right frontal damage and mania or euphoria: Starkstein, S. E., Boston, J. D., & Robinson, R. G. (1988). Mechanisms of mania after brain injury. 12 case reports and review of the literature. *J Nerv Ment Dis*, 176(2), 87–100.

Pathological crying and pathological laughter: Tucker, D. M., Stenslie, C. E., Roth, R. S., & Shearer, S. L. (1981). Right frontal lobe activation and right hemisphere performance. Decrement during a depressed mood. *Arch Gen Psychiatry*, 38(2), 169–174; Sackeim, H. A., Greenberg, M. S., Weiman, A. L., Gur, R. C., Hungerbuhler, J. P., & Geschwind, N. (1982). Hemispheric asymmetry in the expression of positive and negative emotions. Neurologic evidence. *Arch Neurol*, 39(4), 210–218.

Research by Richard Davidson: Davidson, R. (1995). Cerebral Assymetry, Emotion, and Affective Style. In R. Davidson & K. Hugdahl (Eds.), *Brain Assymetry* (pp. 361–388). Cambridge, MA: The MIT Press.

Expression vs. recognition of emotions: Ibid.

Unpleasant or sad images and right-hemispheric activation: Tomarken, A. J., Davidson, R. J., Wheeler, R. E., & Doss, R. C. (1992). Individual differences in anterior brain asymmetry and fundamental dimensions of emotion. *J Pers Soc Psychol*, 62(4), 676–687.

Memory loss and right-frontal activation: Davidson, R. (1995). Cerebral Asymmetry, Emotion, and Affective Style. In R. Davidson & K. Hugdahl (Eds.), *Brain Asymmetry* (pp. 361–388). Cambridge, MA: The MIT Press.

Meditation and left prefrontal activation: Kalb, C. (November 10, 2003). Faith and Healing. *Newsweek, CXLII,* 44–56.

Emotional styles and hemispheric activation: Davidson, R. (1995). Cerebral Asymmetry, Emotion, and Affective Style. In R. Davidson & K. Hugdahl (Eds.), *Brain Asymmetry* (pp. 361–388). Cambridge, MA: The MIT Press.

Sadness and depression when left frontal activation is impaired: Henriques, J. B., & Davidson, R. J. (1991). Left frontal hypoactivation in depression. *J Abnorm Psychol*, 100(4), 535–545.

Right frontal activation in negative emotions: Wheeler, R. E., Davidson, R. J., & Tomarken, A. J. (1993). Frontal brain asymmetry and emotional reactivity: a biological substrate of affective style. *Psychophysiology*, 30(1), 82–89.

Virtual ballgame experiment: Eisenberger, N. I., Lieberman, M. D., & Williams, K. D. (2003). Does rejection hurt? An FMRI study of social exclusion. *Science*, 302(5643), 290–292.

Left and right frontal activation in infants: Davidson, R. J., & Fox, N. A. (1989). Frontal brain asymmetry predicts infants' response to maternal separation. *J Abnorm Psychol*, 98(2), 127–131.

Right and left amygdala activation: Roeder, C., Mueller, J., Sommer, M., Zanella, F., & Linden, D. (2003). Valence But Not Arousal Correlates with Limbic Activity in Emotional Probe Processing in Female Subjects. Paper presented at the Human Brain Mapping, New York City.

Right amydgala and the appreciation of facial expressions of fear: Thomas, K. M., Drevets, W. C., Whalen, P. J., Eccard, C. H., Dahl, R. E., Ryan, N. D., et al. (2001). Amygdala response to facial expressions in children and adults. *Biol Psychiatry*, 49(4), 309–316.

Selecting from restaurant menu: Arana, F. S., Parkinson, J. A., Hinton, E., Holland, A. J., Owen, A. M., & Roberts, A. C. (2003). Dissociable contributions of the human amygdala and orbitofrontal cortex to incentive motivation and goal selection. *J Neurosci*, 23(29), 9632–9638.

Right amygdala and Generalized Anxiety Disorder: De Bellis, M. D., Casey, B. J., Dahl, R. E., Birmaher, B., Williamson, D. E., Thomas, K. M., et al. (2000). A pilot study of amygdala volumes in pediatric generalized anxiety disorder. *Biol Psychiatry*, 48(1), 51–57.

Right amygdala and facial expressions of fear: Anderson, A. K., Spencer, D. D., Fulbright, R. K., & Phelps, E. A. (2000). Contribution of the anteromedial temporal lobes to the evaluation of facial emotion. *Neuropsychology*, 14(4), 526–536.

Brain structures involved in the regulation of emotions: Kolb, B., & Whishaw, I. Q. (1996). *Fundamentals of Human Neuropsychology* (4th ed.). New York: W.H. Freeman.

Lateralization of neurotransmitters (norepinephrine and dopamine): Glick, S. D., Ross, D. A., & Hough, L. B. (1982). Lateral asymmetry of neurotransmitters in human brain. *Brain Res*, 234(1), 53–63.

Dopamine and stereotypic behaviors: Tucker, D. M., & Williamson, P. A. (1984). Asymmetric neural control systems in human self-regulation. *Psychol Rev*, 91(2), 185–215.

Dopamine and addiction: *Ibid.*

Norepinephrine and novelty-seeking: *Ibid.*

Norepinephrine in depression: Delgado, P. L., & Moreno, F. A. (2000). Role of norepinephrine in depression. *J Clin Psychiatry, 61 Suppl 1*, 5–12.

Serotonin in depression: D'Haenen, H., Bossuyt, A., Mertens, J., Bossuyt-Piron, C., Gijsemans, M., & Kaufman, L. (1992). SPECT imaging of serotonin2 receptors in depression. *Psychiatry Res*, 45(4), 227–237.

Kay Redfield Jamison on creativity and psychiatric illness: Jamison, K. (1994). *Touched with Fire: Manic Depressive Illness and the Artistic Temperament.* New York: Free Press Paperbacks; Jamison, K. (1997). *An Unquiet Mind: A Memoir of Moods and Madness.* New York: Vintage Books.

Jablow Hershman and Julian Leib on manic-depressive disorder and creativity: Hershman, D. J., & Leib, J. (1988). *The Key to Genius: Manic-Depression and the Creative Life.* Amherst, NY: Prometheus Books.

Michaelangelo's depression: Hershman, D. J., & Lieb, J. (1998). *Manic Depression and Creativity.* Amherst, NY: Prometheus Books.

Manic-depresive disorder in Napoleon, Hitler and Stalin: Hershman, D. J., & Lieb, J. (1994). *A Brotherhood of Tyrants: Manic-Depression & Absolute Power.* Amherst, NY: Prometheus Books.

Manic-depresive disorder in Potyomkin: Binyon, T. (2003). *Pushkin: A Biography.* New York: Knopf.

Depression as a risk factor for dementia: Roberts, G. W., Leigh, P. N., & Weinberger, D. R. (1993). *Neuropsychiatric Disorders.* London: Wolfe.

Charles Murray's ranking of historic personalities: Murray, C. A. (2003). *Human Accomplishment: The Pursuit of Excellence in the Arts and Sciences, 800 BC to 1950.* New York: HarperCollins.

Connie Strong and Terence Ketter on creative personality: Strong, C., & Ketter, T. (2002, 5/21/2002). *Negative Affective Traits and Openness Have Differential Relationships to Creativity.* Paper presented at the APA Annual Meeting, Philadelphia, PA.

Brain activation patterns in manic and depressed states: Dr. David Silbersweig, personal communication to Elkhonon Goldberg.

Brain activation profile of manic-depressive disorder: Baxter, L. R., Jr., Schwartz, J. M., Phelps, M. E., Mazziotta, J. C., Guze, B. H., Selin, C. E., et al. (1989). Reduction of prefrontal cortex glucose metabolism common to three types of depression. *Arch Gen Psychiatry*, 46(3), 243–250; Delvenne, V., Delecluse, F., Hubain, P. P., Schoutens, A., De Maertelaer, V., & Mendlewicz, J. (1990). Regional cerebral blood flow in patients with affective disorders. *Br J Psychiatry*, 157, 359–365; Migliorelli, R., Starkstein, S. E., Teson, A., de Quiros, G., Vazquez, S., Leiguarda, R., et al. (1993). SPECT findings in patients with primary mania. *J Neuropsychiatry Clin Neurosci*, 5(4), 379–383; Bonne, O., Krausz, Y., Gorfine, M., Karger, H., Gelfin, Y., Shapira, B., et al. (1996). Cerebral hypoperfusion in medication resistant, depressed patients assessed by Tc99m HMPAO SPECT. *J Affect Disord*, 41(3), 163–171.

Subtypes of depression: *Diagnostic and Statistical Manual of Mental Disorder—IV—Text Revision* (4 ed.)(2000). Washington, DC: American Psychiatric Association.

Changes in amygdala activity with age: Leigland, L. A., Schulz, L. E., & Janowsky, J. S. (2004). Age related changes in emotional memory. *Neurobiol Aging*, 25(8), 1117–1124.

13. The Dog Days of Summer

Sulcar shallowing in aging: Rettmann, M., Prince, J., & Resnick, S. (2003). *Analysis of Sulcal Shape Changes Associated with Aging*. Ninth Annual Meeting of the Organization for Human Brain Mapping, New York City.

Insula in aging: Grieve, S., Clark, R., & Gordon, E. (2003). *Brain Volume and Regional Tissue Distribution in 193 Normal Subjects Using Structural MRI: The Effect of Gender, Handedness and Age*. Ninth Annual Meeting of the Organization for Human Brain Mapping, New York City.

Voxel morphometry in aging: Taki, Y., Goto, R., Evans, A., Sato, K., Kinomura, S., Ono, S., et al. (2003). *Voxel Based Morphometry of Age Related Structural Change of Gray Matter for Each Decade in Normal Male Subjects*. Ninth Annual Meeting of the Organization for Human Brain Mapping, New York City.

Brain size reduction in the elderly with depression: Ballmaier, M., Kumar, M., Sowell, E., Thompson, P., Blanton, R., Lavretsky, H., et al. (2003). *Cortical Abnormalities in Elderly Depressed Patients*. Ninth Annual Meeting of the Organization for Human Brain Mapping, New York City.

Age, gender, handedness, and brain volume: Rex, D., & Toga, A. (2003). *Age, Gender, and Handedness Influences on Relative Tissue Volumes in the Human Brain*. Ninth Annual Meeting of the Organization for Human Brain Mapping, New York City.

Changes on WAIS IQ tests with aging: Lezak, M. D., Howieson, D. B., &

Loring, D.W. (2004). *Neuropsychological Assessment* (4 ed.). New York: Oxford University Press. Technically, what declines is not Verbal IQ or Performance IQ per se, but the performance on the tests used to measure them. Then the values are age-corrected to keep the IQ quotients constant.

14. Use Your Brain and Get More of It

Fernando Nottenbohm on neuroplasticity: Nottehbom, F. (1977). Asymetries of neural control of vocalization in the canary. In S. Harnard, R.W. Doty, L. Goldstein, & J. Jaynes, (Eds.), *Lateralization in the Nervous System* (pp. 23–44). New York: Academic Press.

Neuronal proliferation in monkeys: Gould, E., Reeves, A.J., Graziano, M.S., & Gross, C.G. (1999). Neurogenesis in the neocortex of adult primates. *Science*, 286(5439), 548–552.

Neuronal proliferation in hippocampi: Gould, E., & Gross, C.G. (2002). Neurogenesis in adult mammals: some progress and problems. *J Neurosci*, 22(3), 619–623.

Brain structures vulnerable in aging and dementia: Raz, N. (2000). Aging of the brain and its impact on cognitive performance: integration of structural and functional findings. In F. Craik & T. Salthouse (Eds.), *The Handbook of Aging and Cognition* (2nd ed., pp. 1–90). Mahwah, NJ: Lawrence Erlbaum Associates.

Brain derived neurotrophic factor levels: Cotman, C.W., & Berchtold, N.C. (2002). Exercise: a behavioral intervention to enhance brain health and plasticity. *Trends Neurosci*, 25(6), 295–301.

For general review on neuroplasticity: Schwartz, J., & Begley, S. (2002). *The Mind and the Brain: Neuroplasticity and the Power of Mental Force.* New York: Regan Books.

Neurogenesis in human hippocampi: Eriksson, P.S., Perfilieva, E., Bjork-Eriksson, T., Alborn, A.M., Nordborg, C., Peterson, D.A., et al. (1998). Neurogenesis in the adult human hippocampus. *Nat Med*, 4(11), 1313–1317.

Neurogenesis in Alzheimer's disease: Shors, T.J. (2003). Can new neurons replace memories lost? *Science of Aging Knowledge Environment*, 49, 35–38.

Hippocampi in cab drivers: Maguire, E.A., Gadian, D.G., Johnsrude, I.S., Good, C.D., Ashburner, J., Frackowiak, R.S., et al. (2000). Navigation-related structural change in the hippocampi of taxi drivers. *Proc Natl Acad Sci USA*, 97(8), 4398–4403.

Inflammation and neurogenesis: Monje, M.L., Toda, H., & Palmer, T.D. (2003). Inflammatory blockade restores adult hippocampal neurogenesis. *Science*, 302(5651), 1760–1765.

Angular gyrus in bilinguals: Mechelli, A., Noppeney, U., O'Doherty, J., Ashburner, J., & Price, C. (2003). *A Voxel-Based Morphometry Study of Monolinguals, Early Bilinguals and Late Bilinguals.* Ninth Annual Meeting of the Organization for Human Brain Mapping, New York City.

Luria on angular gyrus: Luria, A.R. (1970). *Traumatic Aphasia*: The Hague: Mouton.

Heschl's gyrus in musicians: Schneider, P., Scherg, M., Dosch, H.G., Specht, H.J., Gutschalk, A., & Rupp, A. (2002). Morphology of Heschl's gyrus reflects enhanced activation in the auditory cortex of musicians. *Nat Neurosci*, 5(7), 688–694.

Brain changes in jugglers: Draganski, B., Gaser, C., Busch, V., Schuierer, G., Bogdahn, U., & May, A. (2004). Neuroplasticity: changes in grey matter induced by training. *Nature*, 427(6972), 311–312.

Cell migration in rodents and humans: Sanai, N., Tramontin, A.D., Quinones-Hinojosa, A., Barbaro, N.M., Gupta, N., Kunwar, S., et al. (2004). Unique astrocyte ribbon in adult human brain contains neural stem cells but lacks chain migration. *Nature*, 427(6976), 740–744.

"Immigration denied": Rakic, P. (2004). Neuroscience: immigration denied. *Nature*, 427(6976), 685–686.

Function vs. neuropathology: Katzman, R., et al. (1988). Clinical, pathological, and neurochemical changes in dementia; a subgroup with preserved mental status and numerous neocortical plaques. *Ann Neurol.* 23: 53–59.

School Sisters of Notre Dame: Snowdon, D. (2001). *Aging with grace.* New York: Bantam Books.

15. Pattern Boosters

MacArthur Project: Albert, M.S., Jones, K., Savage, C.R., Berkman, L., Seeman, T., Blazer, D., et al. (1995). Predictors of cognitive change in older persons: MacArthur studies of successful aging. *Psychol Aging*, 10(4), 578–589; Rowe, J., & Kahn, R. (1998). *Successful Aging.* New York: Random House.

Miller on evolutionary origins of art: Miller, G. (2001). *The Mating Mind.* New York: Anchor Books.

Art as "throwaway" activity: *Ibid.*

Beth Neimann on Mozart effect: Personal communication to E. Goldberg.

Memory enhancement programs: Cavallini, E., Pagnin, A., & Vecchi, T. (2003). Aging and everyday memory: the beneficial effect of memory training. *Arch Gerontol Geriatr*, 37(3), 241–257; Ball, K., Berch, D.B., Helmers, K.F., Jobe, J.B., Leveck, M.D., Marsiske, M., et al. (2002). Effects of cognitive training interventions with older adults: a randomized controlled trial. *JAMA*, 288(18), 2271–2281; Rapp, S., Brenes, G., & Marsh, A. P. (2002). Memory enhancement training for older adults with mild cognitive impairment: a preliminary study. *Aging Ment Health*, 6(1), 5–11; Schaie, K. W., & Willis, S. L. (1986). Can decline in adult intellectual functioning be reversed? *Developmental Psychology*, 22(2), 223.

Taub's rehabilitation method: Taub, E., & Morris, D. M. (2001). Constraint-induced movement therapy to enhance recovery after stroke. *Curr Atheroscler Rep*, 3(4), 279–286.

Epilogue: The Price of Wisdom

Sir Edward Dyer's poem: Sargent, R. M. (1968). *The Life and lyrics of Sir Edward Dyer* (formerly entitled *At the Court of Queen Elizabeth*). Oxford: Clarendon P.

INDEX